Digital Transformation in Aviation, Tourism and Hospitality in Southeast Asia

Technological advances and the drive to digitalize business processes in aviation, tourism, and hospitality have forced the industries to go along with the digital movement. The results are often mixed. This book brings together contributions from leading scholars in the field and explores the digital transformation in these industries in Southeast Asia.

The book looks at the impact of digital transformation on the region and the issues and challenges brought about by this transformation. It also addresses trends in the industries from blockchain technology, AI, biometric and mobile technology applications to in-flight catering. It examines the impact of COVID-19 on the industries and how the pandemic has led to businesses adopting new business models. Through the case studies of digital adoptions in the region, readers will gain insights on how the countries have leveraged new technologies and the implementation processes to drive digital transformation.

The book aims to help scholars and policy makers understand the digital advances in the industries to better formulate responses in research and policy making and deliver effective digital transformation.

Azizul Hassan is a member of the Tourism Consultants Network of the UK Tourism Society. He has been working for the tourism industry as a consultant, academic, and researcher for over 20 years.

Nor Aida Abdul Rahman is Associate Professor at Universiti Kuala Lumpur, Malaysia, and currently serves as Head of Aviation Management at Universiti Kuala Lumpur, Malaysian Institute of Aviation Technology, in Subang, Selangor, Malaysia.

Routledge Frontiers of Business Management

The Management of Consumer Co-operatives in Korea
Identity, Participation and Sustainability
Edited by Seungkwon Jang

Entrepreneurship, Innovation and Inequality
Exploring Territorial Dynamics and Development
Edited by Vanessa Ratten, Jose Álvarez-García and Maria de la Cruz del Rio-Rama

Innovation and Industrial Development in China
A Schumpeterian Perspective on China's Economic Transformation
Kaidong Feng

Beauty of Crowdfunding
Blooming Creativity and Innovation in the Digital Era
Sunghan Ryu

Renewable Energy Management in Emerging Economies
Strategies for Growth
Henry K. H. Wang

Statistical Modelling and Sports Business Analytics
Edited by Vanessa Ratten and Ted Hayduk

China's Drive for the Technology Frontier
Indigenous Innovation in the High-Tech Industry
Yin Li

Digital Transformation in Aviation, Tourism and Hospitality in Southeast Asia
Edited by Azizul Hassan and Nor Aida Abdul Rahman

For more information about this series, please visit www.routledge.com/Routledge-Frontiers-of-Business-Management/book-series/RFBM

Digital Transformation in Aviation, Tourism and Hospitality in Southeast Asia

Edited by Azizul Hassan and Nor Aida Abdul Rahman

Routledge
Taylor & Francis Group
LONDON AND NEW YORK

First published 2023
by Routledge
4 Park Square, Milton Park, Abingdon, Oxon OX14 4RN

and by Routledge
605 Third Avenue, New York, NY 10158

Routledge is an imprint of the Taylor & Francis Group, an informa business

British Library Cataloguing-in-Publication Data
A catalogue record for this book is available from the British Library

ISBN: 9781032324654 (hbk)
ISBN: 9781032324661 (pbk)
ISBN: 9781003315179 (ebk)

DOI: 10.4324/9781003315179

Typeset in Galliard
by Deanta Global Publishing Services, Chennai, India

Contents

Figures

Tables

Contributors

Anis Najiha Ahmad is Assistant Professor at the International Institute of Halal Research and Training (INHART), International Islamic University Malaysia (IIUM), Selangor, Malaysia. She graduated from Universiti Sains Malaysia with a bachelor's degree in Food Technology (Hons) in 2008, and an MSc in Food Technology in 2012. Her master thesis, "Development of Chromatography Methods for Determination of Ethanol Halal Limits in Selected Food and Drinks," focused on the development of a routine method for ethanol quantification, and the determination of alcohol limits in halal certification. During her doctorate study, she decided to go in the opposite direction, from the hard-analytical detection to a soft management approach – addressing challenges to ensure the halal and wholesome state of food production. Her PhD thesis focuses on the development of an instrument that assesses factors that lead to effectiveness of halal food management system in SMEs. Her current work focuses on a broad range of topics in the field of halal food quality and safety management, halal consumer behaviour, halal ecosystem, tourism, whistleblowing, and core competencies in the halal industry. She is a certified Professional Junior Trainer for Atlas.ti, one of the most well-known computer-assisted qualitative data analysis software programmes (CAQDAS).

Azlina Mohd Ali is Lecturer of Universiti Kuala Lumpur (UniKL), Kuala Lumpur, Malaysia, with a specialization in aviation management and safety management research. She has more than ten years of experience working in the academic sector and is currently pursuing her PhD degree in management (safety management in aviation). Her research interests include safety management, aviation management, and tourism management. She has actively participated in presenting papers at conferences.

Rita Zaharah Wan-Chik is Senior Lecturer at the Technical Foundation/Aviation Management Section, Universiti Kuala Lumpur – Malaysian Institute of Aviation Technology (UniKL MIAT), Selangor, Malaysia. Previously she taught at the Malaysian Institute of Information Technology (MIIT) under the same university. She received her PhD in Information Studies from the University of Sheffield, UK. She has been in academia for 15 years, mostly teaching information technology-related courses, including the Internet of things (IoT) and data analytics, cyber law, data mining and data warehousing, management information systems, and operating systems, among others. She has also experienced working in the industry for a few years before joining the world of academia. Her research interests include information technology adoption in aviation, information behaviour, and semantic data analytics.

Nur Atiqah Rochin Demong obtained her PhD in Analytics at University of Technology Sydney, Australia. She is Senior Lecturer at the Department of Technology and Supply Chain Management Studies, Faculty of Business and Management, Universiti Teknologi MARA (UiTM), Selangor, Malaysia. Her research interests include big data, data analytics, database, data warehousing, e-commerce, digital entrepreneurship, management of information systems, business analytics, and knowledge discovery. Her contributions have been published in prestigious peer-reviewed journals.

Zuhra Junaida binti Mohamad Husny Hamid is Senior Lecturer at the Department of Urban and Regional Planning, Faculty of Built Environment and Surveying, Universiti Teknologi Malaysia (UTM), Johor, Malaysia. She is also Deputy Director of the Centre for Innovative Planning and Development, UTM. Dr Zuhra completed a PhD and MSc in Transportation Planning from the UTM. She is also a chartered member of the Chartered Institute of Logistics and Transport (CILT), UK, and is registered as a Professional Technologist by the Malaysia Board of Technologies. She has spent more than 15 years in various industries including multinational companies, both local and foreign, among others, Mearsk Line Sdn Bhd and Kuehne-Nagel Sdn. Bhd. Her industrial expertise is on shipping and international logistics arrangement. She has ten years of teaching experience in the subject area of logistics and transportation planning. Apart from that, she is a certified trainer under the Malaysia Halal Professional Board, JAKIM.

Nor Hisham Ghafar is Lecturer at the Malaysian Institute of Aviation Technology, Universiti Kuala Lumpur, a technical university in Malaysia. Ghafar completed his bachelor of Business Administration from the International Islamic University Malaysia and his master of Human Resource Management from the Universiti Utara Malaysia. He is currently pursuing his philosophical doctorate in Management at the Putra Business School, Universiti Putra Malaysia. His research interest is in the areas of human resource management, talent management, aviation management, and academic programme development. Hisham has authored more than ten journal and conference articles as well as book chapters in human resource management, aviation management, and academic development publications. He is also a program developer for a new master's programme in aviation management at his current university.

Mohd Khir Harun is Associate Professor at the Malaysian Institute of Aviation Technology (UniKL MIAT). He has 15 years of experience working in airlines and manufacturing of multinational aerospace company. His research spans a wide range of aviation engineering and management areas including supply chain and transportation. His current interests include IR4.0 in aerospace applications and Urban Air Mobility (UAM) as well as the international travel-related industry.

Siti Salwa binti Hasbullah is Lecturer in the Foundation Centre, Universiti Kuala Lumpur – Malaysian Institute of Information Technology (UniKL MIIT), Kuala Lumpur, Malaysia. She received her Master of Computer Science from the National University of Malaysia. She has been in academia for 11 years teaching information technology (IT) and business-related courses such as computer organization, business intelligence, business enterprise management, and project management. She has also experienced working in the industry for a few years before joining the world of academia. Her current research interests include artificial intelligence, sentiment analysis, natural language processing, information retrieval, and semantic data analytics.

Azizul Hassan is a member of the Tourism Consultants Network of the UK Tourism Society. Dr Hassan has been working for the tourism industry as a consultant, academic, and researcher for over two decades. His research interest areas are technology-supported marketing for tourism and hospitality, immersive technology applications in the tourism and hospitality industry, and technology-influenced marketing suggestions for the sustainable tourism and hospitality industry in developing countries. Dr Hassan has authored over 150 articles and book chapters in leading tourism outlets. He is also part of the editorial team of 25 book projects from Routledge, Springer, CAB International, and Emerald Group Publishing Limited. He is a regular reviewer of a number of international journals.

Sheikh Muhamad Hizam holds the position of Senior Lecturer of Management as well as Dean of Business School, Universiti Kuala Lumpur (UniKL), Kuala Lumpur, Malaysia. He is also involved as the Head of Research Cluster and Editorial Board member for the journal *UBIS*. He formerly worked in the agribusiness, manufacturing, and air logistic sectors dealing with multinational clients like Intel, IBM, Motorola, etc. He specializes in organizational behaviour, human resource management, and stress management. He has over 20 years of work experience in both industry and academia. His area of research is in the different fields of management, i.e., human resource management and technology management, stress management, marketing, and digital HR.She obtained her bachelor's degree in Accounting from Universiti Kebangsaan Malaysia (UKM). She is currently researching accounting, entrepreneurship, and marketing as an effort to expand her knowledge in these fields.

Suhaila Abdul Kadir is Senior Lecturer, Faculty of Entrepreneurship and Business, Universiti Malaysia Kelantan (UMK), Kelantan, Malaysia. She obtained her PhD in Business and Entrepreneurship from the University of Malaya (UM). She has been involved in many research projects and published numerous articles nationally and internationally. She is currently researching entrepreneurship and marketing as an effort to expand her expertise. She also supervises doctoral and master's students and develops teaching modules.

Zawiah Abdul Majid is Senior Lecturer at Universiti Kuala Lumpur and served as Head of Teknoputra, International, Industrial and Institutional Partnerships (3IP), Business School, Universiti Kuala Lumpur (UniKL), Kuala Lumpur, Malaysia. She has 20 years' experience teaching various courses such as logistics, entrepreneurship, and innovation management. Currently, she is a Fellow at the Chartered Institute of Logistics and Transport (CILT), Woman in Logistics and Transport (WiLAT), Global Vice Chairperson (GVC) Southeast Asia, and founder of WiLAT Malaysia. She has more than 30 years' experience in the logistics industry and is currently pursuing her doctorate in halal supply chains. She is on the working group panel of MS2400 Halal Supply Chain for DSM and TC10 for SMIIC, Istanbul, Turkey. She is also chairman of the National Competency Standard (NCS) for Halal Logistics under MOHR, HDC, and JPK, and a certified Halal Professional Board trainer endorsed by JAKIM.

Siti Noorsuriani Maon is Associate Professor at Faculty of Business and Management, Universiti Teknologi MARA (UiTM), Selangor, Malaysia. She received her PhD from the University of Queensland (UQ), Australia, in 2013. She has actively participated in several research works and community projects relevant to her interests. Her research interests include management, behaviour, health informatics, and technological intervention in health improvement.

Ahmad Zahir Mokhtar is a professional licensed aircraft maintenance engineer with more than 20 years of experience in the airline industry. He was the Base Maintenance Manager of Malaysia's national airline responsible for heavy maintenance of narrow-bodied and wide-bodied aircraft such as the Boeing 737 and 747. He has 10 years of experience in various airline corporate services functions, with responsibility for many high-level projects. He graduated from Universiti Teknologi Malaysia with a master's of science in Transportation Planning. He later joined the Universiti Kuala Lumpur (UniKL) as the Dean of the Aviation School, moving up to the Chancellory as the university's Deputy President of Students' Affairs and Industry Relations. He is now a Professor at UniKL Aviation School in Subang Campus focusing his efforts on developing the aerospace teaching factory specializing in helicopter maintenance.

Au Yong Hui Nee is Associate Professor of Economics and Dean of the Faculty of Business and Finance, University Tunku Abdul Rahman, Malaysia. She received her PhD from Universiti Sains Malaysia. She was a Japanese Government MEXT (Monbukagakusho) Scholar. Prior to joining academia, she was trained in an international hotel chain. She held management positions in Fortune 500 multinational corporations and the public sector. She has extensive experience in data analytics and technology management. She has received research grants in the amount of USD150,000. She has published more than 30 papers in Scopus/SSCI-indexed journals. She is a lead editor of a book project from World Scientific Publishing Co. She is a member of the editorial board and reviewer of SSCI-indexed international journals. She is a recipient of the M.I.M. Tun Razak Youth Leadership Award.

Nurhayati Mohd Nur is Senior Lecturer in the Aerospace Department and currently serves as Head of Research and Innovation in Universiti Kuala Lumpur – Malaysian Institute of Aviation Technology (UniKL MIAT). She received her PhD in Engineering (Ergonomics Design) from the University of Malaya. She is the Principal Researcher for the UniKL MIAT Aerospace Research Cluster. Her main research interests include ergonomic design, human factors, manufacturing processes, maintenance activities, and quality management systems. Her research works have been published in reputable academic journals and refereed conference proceedings. Previously, she worked for 12 years in manufacturing industries mainly in the process improvement, quality, and engineering departments. She also worked as a consultant and trainer for ISO 9001, AS 9100, and TS16949 Quality Management System (QMS). She is certified as Lead Auditor for ISO9001, a graduate member of the Board of Engineers Malaysia (BEM), a member of the Chartered Institute of Logistics and Transport Malaysia (CILTM), and an associate member of the Human Factors and Ergonomics Society Malaysia (HFEM) and the Industrial Engineering and Operations Management (IEOM) Society (International).

Emi Normalina Omar obtained her degree in Professional Studies from the Chartered Institute of Logistics and Transport from Universiti Teknologi MARA (UiTM), Selangor, Malaysia. She holds a master's degree in International Logistics from the University of Plymouth (UK) and also she did her PhD in transport and logistics at the Malaysia Institute of Transport (MITRANS). Her PhD thesis is specifically in the area of halal supply chain management. Currently, she is a senior lecturer with more than 15 years' experience and is attached to the Centre for Technology and Supply Chain Management Studies, Faculty of Business and Management, Universiti Teknologi MARA (UiTM), Selangor, Malaysia.

Ravindra Kumar Perumal earned his Doctorate in Management from Malaysia's Limkokwing University of Creative Technology in Cyberjaya. In 2001, he earned a Master of Business Administration degree from Nottingham Trent University in the United Kingdom. Since 2010, he has worked as a senior lecturer at the Asia Pacific University of Technology and Innovation. Prior to that, he worked as a lecturer at Sunway University's School of Hospitality and Tourism Management. For the last 23 years, he has been an academician. In 2012, he was awarded a Fundamental Research Grant Scheme (FRGS) by the Ministry of Higher Education to undertake research on Homestay Tourism as a Medium for Rural Community Development in Malaysia, which he finished successfully in 2014.

Nor Aida Abdul Rahman is Associate Professor, Universiti Kuala Lumpur (UniKL), Kuala Lumpur, Malaysia, and Head of Aviation Management, Universiti Kuala Lumpur – Malaysian Institute of Aviation Technology (UniKL MIAT), Selangor, Malaysia. She has worked as an internal and external trainer in management, supply chain, halal logistics, and postgraduate research. Her research work has appeared in several reputable academic journals such as *Industrial Marketing Management*, *Journal of Humanitarian Logistics and Supply Chain*, *International Journal of Quality and Reliability Management*, *International Journal of Supply Chain Management*, and others. She has also published a number of book chapters and refereed conference proceedings, and is part of the editorial team of a book project with Routledge. She is a panel member of WG in MS2400 Halal Supply Chain standard and TC10 for Halal supply chain standard (SMIIC). She earned a PhD degree in Management (supply chain management) from Brunel University, London, UK. She also serves as an Academic Advisor in college, a chartered member of the Chartered Institute of Logistics and Transport Malaysia (CILTM), an HRDF Certified Trainer, Chairman (Academic Committee) of the Malaysian Association of Transportation, Logistics and Supply Chain Schools (MyATLAS), Vice President (Research Journal) of the Institute for Research in Management and Engineering UK (INRME), a JAKIM Halal Certified Trainer, UniKL Halal Professional Board, and a member of the Academy of Marketing, UK.

NurHusna Samsudin graduated with a BEng (Biochemical-Biotechnology) (Honors) and a second degree in Business Administration from the International Islamic University Malaysia (IIUM) in 2008. Upon graduation, she joined the Biotechnology Engineering Department, IIUM, as a Research Assistant as part of a project entitled "Quantum Chemical Structure Activity Studies of Anticancer Activity of Seconucleoside Nitrosourea Analogs." Later she obtained her MSc (Engineering) from the same university in 2011 under a Biotech Research Grant entitled "Modification and Characterisation of Phytase for Animal Feed Production." Next, she pursued a PhD in Engineering from the Biotechnology Department, IIUM, Malaysia, completed in the year 2017, working on a project entitled "Development of Polycaprolactone Microcarrier with Surface Treatment Using Ultraviolet Ozone System." In 2018, she joined the International Institute for Halal Research and Training (INHART), IIUM, as a Research Fellow for two years and is now working as Assistant Professor at the International Institute of Halal Research and Training (INHART), IIUM, Selangor, Malaysia, starting in July 2020. She is an active researcher in biomaterials, enzyme and molecular modelling, and study on the health benefits of phytochemical compounds particularly from agarwood as an alternative ingredient for halal biopharmaceutical products. She has received a national grant for a project to decipher the prescription in a Malay medical manuscript that contains agarwood as one of the components. She is a member of the Malaysian Society of Bioinformatics and Computational Biology (MaSBiC).

Ilham Sentosa is an ecosystem developer, researcher, senior lecturer, and business consultant with expertise in smart city management, creative technopreneurship, and dynamic business modelling issues using the techniques of system dynamics (SD) and structural equation modelling (SEM). Dr Sentosa is also recognized as an advanced quantitative analyst with an interest in latent variable measurement models (factor analysis, item response), longitudinal data analysis (latent growth curves, growth mixture models), and parametric and non-parametric analysis using measurement models approaches. He holds the post of Associate Professor at the Business School, Universiti Kuala Lumpur (UniKL), Kuala Lumpur, Malaysia, where he teaches a graduate course on advanced structural model analysis techniques.

Farah Hida Sharin is Lecturer at the International University of Malaya-Wales, where she specializes in marketing and entrepreneurship. She graduated with Master of Business Administration (MBA) and Bachelor of Business Administration (Hons.) in Marketing from Universiti Teknologi MARA (UiTM), Selangor, Malaysia. Holding various management positions and being involved in policy development led her to pursue a Doctorate of Philosophy (Management) at UniKL Business School. Prior to joining the teaching profession, she worked in the public and private sectors Her interest in nurturing university graduates directed her to become a basic certified counsellor and a certified HRDF trainer. Her current research interest is in dynamic business modelling using the techniques of system dynamics (SD) and structural equation modelling (SEM).

Reshminder Kaur Satvindar Singh graduated with an MApp in Linguistics from Universiti Putra Malaysia in 2010. She has been lecturing at Universiti Kuala Lumpur – Malaysian Institute of Aviation Technology since 2008. She has experience in curriculum design and development, ESP, language testing and assessment, and interactive learning. She has been actively involved in language training and testing based on ICAO standards for air traffic controllers at Civil Aviation Authority Malaysia (CAAM). She is currently pursuing her PhD with Universiti Malaya, Malaysia. She also provides consultation and short courses for aviation companies and international students. Over the years, she has also been involved in research work and publication in ESP, problem-based learning (PBL), and communication.

Khaeruddin Sudharmin is Visiting Professor at DRB HICOM University of Automotive Malaysia (DHU), Pahang, Malaysia; Adjunct Professor at UNITAR International University, Selangor, Malaysia; and Senior Research Fellow at Asia e University (AeU), Selangor, Malaysia. Sudharmin is a hands-on corporate professor with over four decades of corporate/GLC credentials with exposure to a wide range of industries including publishing, university and investment cooperatives, travel, ICT, fund management, and automotive insurance. His last post before retirement was MD/CEO of MRC Malaysia in partnership with THATCHAM UK. He is a recipient of the Asia HRD Award 2013 (Jakarta) and Best Blue Ocean CEO Award 2011 (Bali).

Mohd. Iskandar bin Illyas Tan is Senior Lecturer at the Department of Information Science, Azaman Hashim Business School, Universiti Teknologi Malaysia (UTM), Johor, Malaysia. Dr Iskandar completed a PhD and MSc in Information Science from the UTM. He is also a registered as a Professional Technologist by the Malaysia Board of Technologies. He holds a Doctor of Philosophy Degree in Information Science from the Faculty of Computing; due to his interest in entrepreneurship and frequent involvement in such activities, he co-founded a company called HOLISTICS Lab Sdn Bhd, a UTM spin-off company focusing on the commercialization of technology that helps the halal industry. One of the innovations that has been developed is QuikHalal, the world's first "cloud services"-based halal auditing application. Dr Iskandar has received various awards and grants on said innovation and has been recognized by various agencies for his innovations.

Fernando Mayor-Vitoria is an industrial engineer and has PhD in economics from the Universitat Politecnica de Valencia and he also holds an MBA and a master's in textile engineering. He has more than 20 years of international business experience in different sectors such as textiles, footwear, and lighting. Moreover, he is Professor at the Universitat Politècnica de València, València, Spain, in the areas of economics and business administration. He has published different papers related to the fields of logistics, big data, and decision science as well as books and chapters about operational management and business strategy.

Tri Wahyuningsih completed his Bachelor of Economics at Wijaya Kusuma Purwokerto University and Master of Management at Sebelas Maret University. Wahyuningsih is Senior Lecturer, Mandala Bhakti Surakarta Tourism Academy, Central Java, Indonesia. Wahyuningsih is a rural tourism activist, who has succeeded in assisting national-level for two rural tourism including Sumberbulu rural tourism and Lembah Dongde rural tourism. Wahyuningsih is also the chairman of a foundation that organizes education for vocational high schools.

Putu Ngurah Suyatna Yasa is Associate Professor at Warmadewa University, Bali, Indonesia, who currently serves as secretary of the postgraduate programme. In addition, his daily duties are as a teaching lecturer in undergraduate and postgraduate programmes. In the undergraduate programme, the subjects taught are econometrics, statistics, microeconomics, macroeconomics, and managerial economics. Meanwhile, in the master's programme, the subjects taught are research methods and quantitative analysis related to the use of structural equation modelling (SEM) for students, as well as marketing management research, human resource management, and multivariate analysis in financial research.

Fathien Azuien Yusriza graduated from Universiti Kuala Lumpur – Malaysian Institute of Aviation Technology (UniKL MIAT), Subang, Malaysia, with a bachelor's degree in Aviation Management. She has experienced working with global logistics companies and is currently pursuing her master's degree at Universiti Kuala Lumpur – Malaysian Institute of Aviation Technology (UniKL MIAT), Selangor, Malaysia. She is working on her research related to the effectiveness of aviation's supply chain management operation. Her interest in pursuing research relating to aviation's supply chain began from the day she completed her undergraduate thesis on the effectiveness of inventory management in airline operations. In the year 2022, she was given the opportunity to participate in an exchange mobility programme at a prestigious university in Italy called Politechnico Di Torino (POLITO) for engineering management courses. She will complete her master's degree by 2023 and aspires to serve and bring meaningful conversations to the table that will benefit the aviation industry as a whole.

Introduction

Azizul Hassan and Nor Aida Abdul Rahman

Aviation, tourism, and hospitality in Southeast Asian countries arguably have been experiencing digital transformation over the years. In this particular region of the world, digital transformation aims to satisfy the changing aviation, tourism, and hospitality business and market requirements. It also entails employing the most recent technologies for either adapting the current or developing new culture, processes of business, and consumer experiences. Technological innovations in this region are widely applied in both the consumer and business markets of aviation, tourism, and hospitality as one of the key factors to benefit these industries, and to improve both the tourists' and travellers' experiences.

Literally, technology stays by the tourists' and travellers' side during the entire purchase journey. One of the most important technologies in aviation, tourism, and hospitality is mobile technology. The cell phone has become the tourists' and travellers' tour guide, communicates with travel agencies, books airline tickets, finds the best restaurants, acts as a map, and so on. The Internet of things (IoT) plays a significant role in aviation, tourism, and hospitality. On the other hand, IoT brings significant updates to these industries with the support of integrated sensors connected to the Internet inside items like cars, suitcases, buildings, and more.

The technological characteristics of aviation, for example, show that the adoption of the emerging technologies helps ensure the safety of travellers. Globally, aviation is recognized as a highly regulated industry that helps the economic growth of any country. Aviation supports many other industries such as tourism, manufacturing, business, etc. Also, technology plays a significant role in aviation as most travellers use aircraft to travel from one location to another.

In tourism, technological innovations, tourism technology, or tourism automation benefit both the tourists and the travellers. Technological innovations, such as flight monitoring, online travel and tourism firms, and systems, allow tourists and travellers to rate their experiences. These also offer significant benefits allowing tourist to enjoy their travel and stay connected with the family. As far as technology advances in tourism are concerned, smartphones, smart speakers, and artificial intelligence (AI) assistants are all supporting tourists and travellers. Voice search is increasingly being used by tourists and travellers to check and get airline tickets, hotel rooms, and travel experiences.

DOI: 10.4324/9781003315179-1

Hospitality is competitive and essentially needs to keep up with the very latest technological trends. Technology can help hospitality to streamline processes and to improve the level of customer experience. Technology supports hospitality, especially with regard to the widespread adoption of voice search, virtual reality (VR), augmented reality (AR), AI, the IoT, fingerprint recognition, and so on. Contactless payment for instance has become more popular in hospitality settings, while mobile check-ins have emerged in hotels, restaurants, and airports. Many of these technologies have grown because they help to reduce friction, and cut waiting times for hospitality customers.

Aviation, tourism, and hospitality still require more research studies to analyse the trend of digital transformation. Thus, the call to study digital transformation in aviation, tourism, and hospitality is a timely demand as these are among the fastest growing industries in Southeast Asia contributing to socio-economic development. On this note, this book focuses on the features, breadth, and diversity of digital transformation in aviation, tourism, and hospitality in Southeast Asia.

Hamid and Tan present the context of digital transformation in Southeast Asian aviation, tourism, and hospitality in the first chapter. In the recent era, digital transformation has not only enhanced, but also revolutionized or disrupted every company function. These innovations, which are now being used in aviation, tourism, and hospitality, will change the way customers, industries, and businesses operate. Southeast Asia is rapidly growing in the areas of aviation, tourism, and hospitality, all of which are important economic drivers. In this chapter, we look at how digital revolution is affecting aviation, tourism, and hospitality in Southeast Asia. A thorough examination of Internet marketing, computer systems, mobile communication, technology trends, AI technology, AR technology, VR technology, the IoT, voice recognition technology, and many other topics is conducted.

Nee and Dean, in Chapter 2, discuss the Association of Southeast Asian Nations (ASEAN) region's digital transformation influences, issues, and challenges in aviation, tourism, and hospitality.

Ahmad and Samsudin look at the theoretical and practical aspects of blockchain technology in aviation, tourism, and hospitality in the third chapter. The chapter covers blockchain's essential underlying technology and discusses how it might be used in the tourist industry, which encompasses the aviation and hospitality industries. The most significant obstacles to blockchain adoption are also explored.

Harun, Mokhtar, Singh, and Rahman highlight how artificial intelligence, biometric technology, and mobile technology application in aviation and tourism are changing the face of tourism and air travel in the fourth chapter. In this chapter, the researchers will explain how the current developments in three technologies, namely AI, biometrics, and mobile technology, have revolutionized the landscape of the global air travel and tourist business.

In the fifth chapter, Ghafar discusses present and prospective big data analytics advances in the Southeast Asian aviation, tourist, and hospitality industries. This chapter also seeks to give a detailed explanation of big data analytics, with an emphasis on the technology's recent developments and important problems.

In the sixth chapter, Yusriza and Rahman examine the relationship between technology and in-flight food in air travel, tourism, and hospitality. The primary goal of this chapter is to analyse the role of technology in the airline catering industry and to explain how technology applications in the airline catering industry are related to air travel, tourism, and hospitality. Second, this chapter discusses present and prospective technology applications in airline catering, and third, this chapter examines case studies of many airline caterers who use technology. This chapter concludes with a discussion of existing research and opportunities for future research.

Kadir and Hussin study cross-country instances of technology use in the Asian tourist and aviation industries in the seventh chapter. The purpose of this research is to examine the technology applications of tourism and aviation industry growth in Malaysia and Thailand, two of the most populous ASEAN countries. The link between smartphone technology and the tourist and aviation sectors in Malaysia and Thailand is also examined in this study. TripAdvisor, PlateCulture, Offpeak, chatter bus, and KL Transit are just a few of the smartphone apps that may help tourists organize their travels better. The study also investigates the Internet's role as a source of information in the tourist and aviation industries.

In the eighth chapter, Sharin, Sentosa, and Perumal investigate whether Malaysia's resilient model for rural tourism sustainability is either chaotic or orderly. The tourist industry has been hit the hardest by the crisis. As a result, this research highlights the role of rural tourist activities in rural regions' economic growth in Malaysia before and during the COVID-19 pandemic, as well as providing an analysis of rural tourism locations. As a consequence, an integrated empirical model for rural tourist sustainability (RTS) was suggested, with community resource tourism (CRT) as a causative component and technical development processes (TDP) as a mediator.

Wan-Chik and Hasbullah look at the current and future advancements of big data analytics in Malaysia's aviation, tourism, and hospitality industries in the ninth chapter. This chapter provides an overview of big data and data analytics in the aviation, tourism, and hospitality industries, as well as their present and future developments. In the aviation, tourism, and hospitality industries, a study of previous and present advancements in the usage of big data and data analytics was conducted. The adoption of data analytics by various stakeholders in certain sectors, such as authorities, decision makers, domain researchers, application users, and typical consumers, is discussed in this chapter.

Yasa and Sentosa present an empirical study on the technology usage dimensions within the tourism craft business in Bali, Indonesia, using a structural equation modelling technique in the tenth chapter. Using a quantitative technique, the current study created a connection between technology usage and company growth in the dynamic movement of the tourism craft sector. To determine the three aspects of technology usage that were postulated, a second order confirmatory factor analysis (CFA) was used.

Wahyuningsih, Sentosa, and Hizam examine technological policy guidance on rural tourism sustainability in Indonesia in the 11th chapter. This research proposes technological policy recommendations for the long-term viability of rural tourism in Indonesia.

Omar, Demong, and Maon wrote the 12th chapter, which analyses the tourist industry during the COVID-19 pandemic and proposes the Tourism Industry Technology Framework (TITEF). This research recommended the TITEF as a way to improve the tourism industry during the COVID-19 pandemic.

Mayor-Vitoria discusses the necessity of contactless hospitality technology in the post-COVID-19 age in the 13th chapter, based on the preceding discussion. This technological acceleration is already laying down a trail and establishing trends that we cannot ignore in the post-COVID-19 age. This chapter also examines a number of factors that are required for the long-term viability of tourist firms and must be considered while rethinking tourism plans.

Again, Mayor-Vitoria focuses on virtual tourism and rebuilding tourist business in the post-COVID-19 period in the 14th chapter. The current situation, which has arisen as a result of the COVID-19 health crisis, is fraught with uncertainty and offers a slew of issues for society in general and tourism in particular. This situation has resulted in the temporary absence of some of the classic tourism-related sustainability issues in select tourist destinations, which the scientific community has viewed as a once-in-a-lifetime chance to rethink tourism and develop a new model based on sustainability. This chapter will look at how an unprecedented worldwide crisis is affecting the tourist industry and how it is being addressed.

Ali, Rahman, and Hassan discuss the history, and present and future trends in technology as well as the safety of aviation and tourism in the 15th chapter. The goal of this study is to close the gap by elucidating the correlational link between safety and technology in aviation and tourism. This chapter expands on the topic of safety in aviation and tourism by emphasizing the significance of maintaining traveller safety throughout their tourist activity or trip from one place to another. This study focuses on the role of technology in ensuring the safety of travellers during air travel and tourist activities.

In Chapter 16, Rahman, Hassan, Sudharmin, Majid, and Nur discuss the future of technological research in aviation, tourism, and hospitality events. First, this chapter will discuss the importance of technology in the aviation industry and the tourism sector, with a focus on tourism event activities; second, it will highlight future technology applications for tourism events; and third, it will explore future research areas in the area of technology and tourism events.

In the 17th chapter, Rahman and Hassan discuss the development and future research priorities for digital transformation in the Southeast Asian countries' aviation, tourism, and hospitality industries. This chapter also discusses the necessity of digital transformation in the aviation, tourist, and hospitality industries in responding to technological, managerial, and marketing challenges.

This book delves deeply into Southeast Asia's digital revolutions in tourism, hospitality, and aviation. This book is deepened with in-depth knowledge about digital transformation technologies in tourism, hospitality, and aviation, as well as a solid understanding of related concepts. This book introduces ideas and dialogues that readers will find extremely beneficial.

Part 1

The Contexts

1 The Context of Digital Transformation in Aviation, Tourism, and Hospitality in Southeast Asia

Zuhra Junaida binti Mohamad Husny Hamid and Mohd. Iskandar bin Illyas Tan

Introduction

Throughout the last decade Southeast Asia has undergone remarkable growth in international tourism. In 2019, travel and tourism contributed US$8.9 trillion to the global gross domestic product (GDP) (10.3%). The travel and tourism industry has generated 330 million jobs (1 in 10 jobs) worldwide. According to the World Travel Tourism Council (WTTC) (2020, Southeast Asian tourism has generated US$380 billion which is 12.1% of world travel and tourism performance. Tourist arrivals and receipts in this region grew at a faster rate compared to most parts of the world and maintained a fourth place among the strongest growth regions at 4.6%. This projection shows that Southeast Asia's tourism industry has developed over the succeeding decade, and plays a key role in the region's total GDP (Hall and Page, 2000). Due to this tourism is deeply embedded in the country's national development policies. Figure 1.1 shows the Asia Pacific's 2019 tourism and travel contribution to GDP (in billion) and the percentage (%) of its GDP growth.

Aviation is one of the main enablers of the tourism and hospitality industry. Aviation provides an essential connection to the fast-growing global tourism industry. The aviation and tourism industries are interdependent for sustainable growth, as it is estimated that more than half (58%) of all foreign tourists travel by air. Aviation's contribution to tourism employment and GDP is also very important. Including direct, indirect, and induced impacts, aviation supports more than 44.8 million jobs within tourism and contributes approximately US$1 trillion annually to global GDP (Aviation Benefits Beyond Borders, 2022). This includes hotels, restaurants, attractions spots, land and air transportation, etc.

In the recent era, rapid technological advancement has not only improved but also changed and disrupted many business operations. Disruptive technology is an innovation that revolutionizes the way consumers, industries, and businesses work. It has also been exploited or used in the travel and tourism industry. The new era of technological revolution is known as the Fourth Industrial Revolution (4IR). The previous Industrial Revolution (IR) already greatly impacted industry, businesses, and society as a whole. With so many new innovations and technological

DOI: 10.4324/9781003315179-3

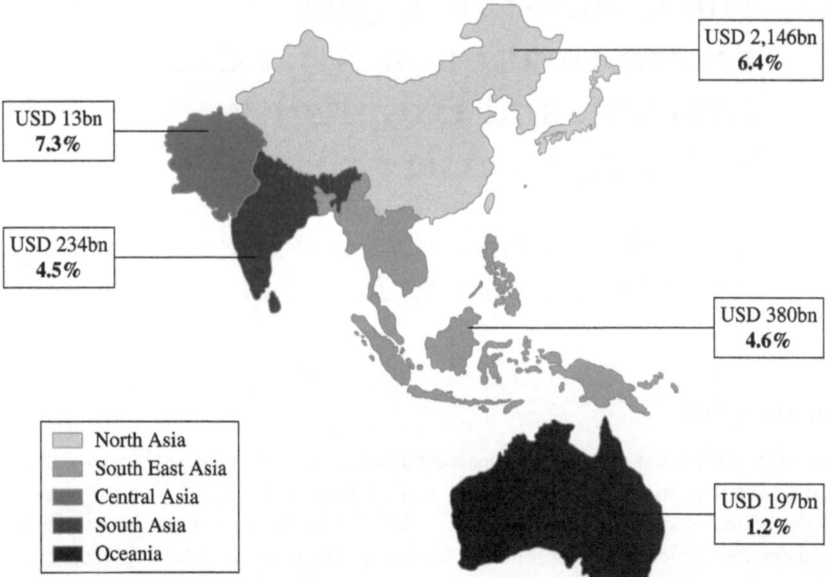

Figure 1.1 Asia Pacific's 2019 tourism and travel contribution to GDP (in billions) and GDP growth percentage (%) (source: World Travel Tourism Council, 2019).

advances in the area of information and communication technology (ICT) and, for example, Internet of Things (IoT), artificial intelligence (AI), autonomous vehicles, robotics, and many more, we have already seen the application of these innovative solutions in developed countries.

Among the technologies that have highly impacted the tourism and hospitality industry is ICT. Thus, ICT has played a vital role in this industry over the last decade. ICT has facilitated reducing costs, enhancing efficiency, and improving the customer experience. Both companies and travellers are able to benefit from better-quality communication, making reservations through guest service systems. ICT has helped the industries in replacing expensive manpower with technology. This helps not only in reducing labour costs, but also helps in avoiding customer service issues. Although in South Asian countries the tourism and hospitality industry is quite slow in responding to the latest innovation, several digital projects have been implemented and positive results have been reaped from those initiatives. This article will discuss the background and context of technology application in the Southeast Asian travel and tourism industry.

Southeast Asia's Aviation

The Southeast Asian aviation market is projected to register a compound annual growth rate (CAGR) of more than 4.5% for the forecast period from 2018 to

2027 (Mordor Intelligent, 2022). The modest increase in passenger transport is driving airline procurement and expansion plans in the region. With support from the government, Southeast Asian countries are also enhancing their aviation infrastructure to support future growth. However, during the COVID-19 pandemic which saw global movement control orders (MCO), countries such as Vietnam, Cambodia, the Philippines, and Thailand reduced air passenger numbers by 34%, 34%, 36%, and 40%, respectively. This reduction in passenger numbers and load factors may impact airline revenues and impact future expansion and procurement plans over the next few years.

According to the report by Mordor Intelligent (2022), the commercial aircraft segment in Southeast Asia held the highest market share in 2019 due to the procurement of new aircraft by airline companies to deal with increasing air passenger traffic. The increase in air transport demand was mainly due to the increase in disposable income levels of people globally as well as in the region and the growth of the tourism and hospitality sector in various countries in Southeast Asia. In the coming years, Thai Airways intends to retire some Boeing 747-400s and Boeing 777s, replacing them with 15 newer-generation aircraft. Singapore Airlines plans to retire 10 Airbus A330-300s, 7 Boeing 777-200s, and 4 Boeing 777-200ERs, while introducing 16 Airbus A350-900s and 6 Boeing 787-10s. As of February 2020, about 150 aircraft were scheduled to be delivered during the forecast period (2018–2027) to accommodate the increase of 16.35% in air passenger traffic at Changi Airport during 2016–2019. This has made Changi Airport one of the busiest airports in the world.

Southeast Asia's Tourism and Hospitality Industry

The tourism and hospitality industry has become an important part of life as it is a mechanism for people to refresh their minds and lower stress, a way to relax and unwind and get away from the hectic hustle and bustle of the city. Tourist infrastructure has developed greatly, as has the variety in the tourism and hospitality industry. The diversity of tourism according to travellers' preferences ensures that there is something for everybody to enjoy and experience. There are 63 categories of tourism. Different types of tourism cater to different travellers' travel demands and behaviours. Among these types of tourism, 18 types are the most relevant to the Southeast Asian tourism industry. In this section, not only can travellers take this information to properly plan their next journey, but this information can also be used by tour operators or travel agencies and governments in building up strategies and expanding the tourism and hospitality industry in Southeast Asia.

Adventure Tourism

Adventure tourism is a type of tourism that involves the exploration of beautiful remote places where travellers can expect the unexpected. It not only involves connecting with a new culture or place but also being active physically. This tourism type does spend more time outdoors and on adventure activities such as rock climbing, kayaking, deep water soloing and river rafting, scuba diving, hiking and trekking, ziplining, canyoning, surfing, sandboarding, and base/bungee jumping. Among the popular places to go for kayaking are Vietnam: Cat Ba and Ha Long Bay, Cambodia: Tonle Sap Lake, Laos: Nam Khan River, Myanmar: Inle Lake, and Thailand: Phang Nga Bay. As for river rafting, popular destinations include Bali, Indonesia: Ayung River, Suwalesi, Indonesia: Ranoyapo and Nimanga Rivers, Malaysian Borneo: Padas River, and Northern Thailand: Pai River. Other forms of adventure tourism are rock climbing and deep water soloing. The well-known places for these activities in Southeast Asia are Thailand: Railay and Ton Sai, Vietnam: Ha Long Bay and Cat Ba, Laos: Vang Vieng and Luang Prabang, and Thailand: Chiang Mai. For scuba diving, Thailand: Koh Tao, Thailand: Koh Phi, the Philippines: Malapasuca, the Philippines: Moalboal and Apo Reef, and Indonesia: Gili Islands are perfect destinations. Hiking and jungle trekking activity is perfect when you can do it with spectacular views of rice fields, or even going up to active volcanoes in Southeast Asia. The famous places to visit for this are in East Java; popular climbs start from Yogykarta and head to the summit of Mount Semeru or Mount Bromo. For an easier climb, try Bali's Gunung Batur, Gunung Agung, and Gunung Rinjani which take one to three days to climb. Among others, popular mountains for this activity are Malaysia: Mount Kinabalu, Laos: Luang Prabang, Vietnam: Mount Fansipan, Sapa valley, and Thailand: Pai and Chiang Mai in Northern Thailand. Thailand: Chiang Mai, Koh Phangan, and Krabi, the Philippines: Lake Sebu and Boracay, Singapore, and Malaysia: Bukit Tinggi offer great ziplining fun and adventure destinations. For canyoning, and adventure that combines swimming, hiking, rock slides, and cliff jumping, Cebu Island, the Philippines, and Dalat in Vietnam are good destinations. In Indonesia, Bali is the place to visit. Surfing is also a major activity for tourists in Southeast Asia as the majority of its countries are surrounded by ocean and has beautiful beaches. The Philippines: Siargao, and Indonesia: Bali are the best places to go for surfing. Sandboarding is quite a new adventure activity in Southeast Asia. The best sandboarding sites in Southeast Asia are Ilocos and Norte in the Philippines and Norte and Mui Ne in Vietnam. Last but not least, there are base and bungee jumping adventures. Singapore: GMAX Reverse Bungee, Thailand: Pattaya, and Malaysia: Kuala Lumpur Twin Tower KLCC offer great experiences for this.

Agritourism

Agritourism, also known as "agrotourism," usually happens in the countryside, particularly farms. It provides travellers with the opportunity to experience rural or village life, taste the local-made food, and get hands-on experience with various farming tasks. According to the Southeast Asian Regional Centre for Graduate Study and Research in Agriculture (2019), in Southeast Asia, for this type of tourism, the Philippines is among the top countries in the world that provide farm tourism locations due to its rich natural resources, rich farming heritage, biodiversity, and warm hospitality. Famous farm tourism locations in the country include La Union, Laguna, and Davao. International tourist arrivals recorded for agritourism from January to May 2018 came to 3.177 million, a 10.24% increase from the same time period of the previous year. The majority of visitors come from South Korea (22.2%), Japan (8.68%), Canada (3.79%), and Taiwan (3.19%). Agritourism includes direct market agritourism, experience and education agritourism, and event and recreation agritourism.

Business Tourism

Business tourism is a type of tourism that involves travelling to a destination away from the usual work place for business matters; the expenses are covered by the company. It is a type of travel for professional purposes. Among the types of business tourism are attending exhibitions and trade fairs, conferences and meetings, and also corporate events. Singapore is 1 of the top 25 destinations for business tourism. Singapore is known as Southeast Asia's economic driving force. Its city-state harbours are among the best business growth environments worldwide, and it is also a major financial capital (Ringgins, 2020). Singapore has the capacity and capability to provide a wide range of advanced industries, flagship conferences, and an abundance of modern-day architectural surprises.

Cultural Tourism

The type of tourism where travellers engage with another country's or its religion's culture is called cultural tourism. Due to its lifestyle and culture that formed people's ways of life, Southeast Asia is one of the most fascinating and culturally complex regions in the world (King, 2012). Tourists are submerged in activities related to rituals and festivals. Asian tourism has changed the way people perceive the cultural details, including heritage sites, of tourism development (Winter et al., 2009). Twenty out of the 31 inscribed United Nations Educational, Scientific and Cultural Organization (UNESCO) cultural and natural sites, across the majority of Southeast Asia countries including Thailand, Malaysia, Vietnam, Cambodia, Laos, the Philippines, and Indonesia. The UNESCO sites in Malaysia are Melaka, George Town in Penang, and Mulu and Kinabalu in Sabah.

Domestic Tourism

Domestic tourism is when residents of a country travel to spend their vacation within the same country. Due to the current situation with the coronavirus pandemic crisis, Southeast Asia, being a world-famous tourism destination, has definitely felt the pinch. Most Association of Southeast Asian Nations (ASEAN) countries have forbidden international flights. Thus, this has severely affected the region's tourism industry. Due to this, ASEAN countries have started to promote "travel local campaigns." Malaysia has launched a promotional travel deals to promote domestic tourism. Eighty thousand flight seats were sold between 10 and 28 June, travelling between July and September 2021 through AirAsia's domestic travel promotions. The same is true of Thailand. The Thai government has approved a domestic tourism motivation promotion package valued at US$718 million to expedite recovery of the country's tourism industry, while Cambodia accepted almost 400,000 local tourists during the first 3 weeks of May 2020 (Nortajuddin, 2020).

Escorted Tours

A type of tourism where travellers are guided in groups is called escorted tourism. Among the usual activities are sight-seeing tours where a tourist guide leads the group of travellers to cultural or historical spots. The tourist guide or tourism agency will handle all the planning for the tour. Travellers can sit back and relax while enjoying the trip, knowing that the tourist guides/agencies are providing the best tour plan. Most countries in Southeast Asia offer escorted tourism packages at a very affordable price. Among the most popular destinations are Vietnam, Cambodia, Thailand, Malaysia, and Indonesia.

Experiential Travel

Experiential travel, also known as immersion travel, is a type of tourism in which travellers experience the destination country through engaging with the country's history, culture and customs, people, and also its cuisine. Among the famous experiential immersion locations in Southeast Asia are home staying in a Thai village, meeting orang-utans in Borneo, climbing active volcanoes in the Tengger Caldera, Indonesia, cruising in a boat along the colourful floating market in Kanchanaburi, trekking through local villages to get to Inle Lake, Myanmar, and scaling the tallest mountain in Southeast Asia to name a few.

Ecotourism

Another type of tourism involving touring natural habitats in a fashion designed to minimize ecological impact is known as ecotourism. The activities comprise visiting untouched natural sites. It encourages the effort of safe-guarding the environment and enhancing the well-being of residents. Southeast Asia is one of the key players in ecotourism. Millions of travellers around the world flock to the region for its beautiful people, islands and beaches, vibrant culture, delightful cuisines, ancient religious architecture, lush jungle, and rich nature sites. Among the most popular sites in Southeast Asia in 2020 and 2021 according to the Trip Discover website are Kho Kong, Cambodia; Luang Namtha, Laos; Taman Negara National Park, Malaysia; Donsol and Peleliu, the Philippines; and Way Kambas National Park, Indonesia.

Fashion Tourism

An arising market segment that evolved from shopping tourism, creative tourism, and cultural tourism is fashion tourism. The activities involved in this type of tourism include visiting a particular place to appreciate, try out, uncover, learn about, buy, and sell fashion. The top fashion destinations in Southeast Asia are Hong Kong, Singapore, and Thailand. Hong Kong is currently one of the top tourist destinations for fashion and is known as the Asian's fashion capital. Hong Kong, the modern minimalist trendsetter, has become an export country and the world's largest supplier of fashion products, specifically clothing and accessories. In promoting fashion tourism in Singapore, the Singapore Tourism Board has introduced Singapore Fashion Week which is a prestigious element in boosting the country's destination attractiveness (Singapore Tourism Board, n.d.). Thailand is another destination for fashion tourism. Thailand received an award as "The Best Destination for FIT" at the 2018 – National Tourism Fashion Awards. Thailand is famous for its luxurious Thai silks and rich and colourful textiles. However, in Bangkok, low-priced fashionable clothes are abundant. Young Thai designers are able to produce hip, trendy, and stylish designs that have begun to be recognized in the international market.

Halal Tourism or Muslim-Friendly Hospitality

Halal tourism or Muslim-friendly hospitality is an effort to cater to the needs of Muslim visitors. The facilities and services provided must be able to accommodate Muslim visitors' requirements and be in accordance with Shariah law. Among the requirements of Muslim travellers are accommodation which is alcohol-free and separate or dedicated spa facilities and swimming pools for men and women. Malaysia is not only a famous Halal tourism or Muslim-friendly hospitality destination in Southeast Asia but also in the world, followed by Saudi Arabia, the UAE, Morocco, Turkey, Jordan, Qatar, and Indonesia. This is to be expected because Malaysia has already developed its own standard known as the Muslim Friendly Hospitality Standard MS2610:2015. This standard has been well adapted by tourism business operators such as hoteliers, tour guides, and tour package providers to ensure Muslim visitors are well taken care of. Halal tourism is another huge source of potential for the tourism industry. The number of Muslim tourists is expected to grow to a total of 180 million visitors by 2020, generating US$212 billion from 116 million visitors in 2014 (Standing Committee for Economic and Commercial Cooperation of the Organization of Islamic Cooperation, 2016). This number reflects the demands for Halal tourism in various categories of businesses such as hotels, restaurants, and food premises.

Couple Retreats or Honeymoon Tourism

Couple retreats or honeymoon tourism is usually designed for a couple getaway, either for married or non-married couples. Generally, these couples spend more money on this than any other trip. Southeast Asia offers quite a number of honeymoon destinations. Travellers can choose the destination depending on what they look for on their visit, whether for pampering and relaxation, all-day sightseeing, or action-packed adventures. Popular honeymoon destinations in

Southeast Asia are the Khao Sok National Park, Southern Thailand; Bohol and Palawan in the Philippines; Siem Reap and Kep in Cambodia; Hanoi and Ho Chi Minh City (Saigon) in Vietnam; and Mekong in Laos to name a few.

International Tourism

Globalization has caused international tourism to increase. According to the US Department of Economic and Social Affairs (2010), international tourism comprises inbound tourism and outbound tourism, that is, the activities of resident visitors outside the country of reference, either as part of domestic or outbound tourism trips, and the activities of non-resident visitors within the country of reference on inbound tourism trips. In Southeast Asia, most travellers come from the western countries. These visitors are looking to experience a different culture, environment, climate, and food from what they usually have. Thailand and Vietnam are the two most popular destinations.

International Volunteering Tourism

This type of tourism is also known as volunteerism. The travellers are usually people working voluntarily for an organization such as a non-government organization (NGOs) or a special cause such as a rescue mission outside their home country. This kind of tourism has turned out to be trendy in many countries. In this tourism, travellers and volunteers are able to provide help to local people in different forms, from teaching English to schoolchildren, to helping preserve endangered animals such as elephants and orang-utans. This also attracts ecotourism; travellers may even help to build a community centre or repair a religious building. This type of tourism is popular because the volunteers are able to enrich their souls by channelling their energy and enthusiasm into doing tasks helping the less fortunate. Among the famous international tourism destinations of the world are the Philippines, Thailand, Cambodia, Indonesia, and also Malaysia.

LGBT Tourism

According to the United Nations World Tourism Organization (2017), this tourism implies the "development of marketing of tourism products and services to lesbian, gay, bisexual or transgender people." The ASEAN Human Rights Declaration 2012 does not include any provision on LGBT rights or corresponded with the general attitude of Southeast Asian countries towards acceptance of LGBT. However, in the year of 2011, the Philippines and Thailand supported a joint statement on ending acts of violence related to human rights violations based on gender identity and sexual orientation. This has encouraged the LGBT tourism industry in those countries.

Libel Tourism

Libel tourism is where travellers who are pursuing a lawsuit seek foreign country jurisdiction where the laws of defamation are considerably more plaintiff-friendly.

Usually travellers stay in that foreign country until the case is closed. Singapore is among the countries that are popular legal tourism destinations.

Medical Tourism

Medical tourism involves travelling to another country to obtain advanced or more reasonably priced medical treatment. Usually patients from developed countries come to a developing country to gain access to affordable medical care. In most Southeast Asian countries, medical tourism makes up at least one-third of private hospitals' income. Thailand, Malaysia, and Singapore are the leading countries in the field of medical tourism in ASEAN. Nevertheless, the Philippines and Vietnam are also a growing market offering relatively decent quality medical care at a very affordable cost, compared to other neighbouring countries.

Self-Guided Tours/Backpack Tourism

Self-guided tours, also known as solo trips or as backpack tourism, are another type of tourism where the traveller is in charge of moving from one point to another themselves. The traveller plans the trip themselves and travels based on data and information such as location maps, travel guides, and recommendations provided by other tourists who have previously gone to the place. Southeast Asia is a great destination for backpackers. The places are easy to travel around, safe, and there are other travellers around. Some of the popular sites in Southeast Asia are Singapore, Vietnam (Mekong Delta), the Philippines (Boracay, Cebu Island), Thailand (Phuket to Bangkok), Indonesia (Bali, Yogyakarta), and Malaysia.

Wellness Tourism

Wellness tourism concerns travelling to nurture health through activities related to physical, psychological, or spiritual health. Unlike medical tourism, wellness tourists travel with the aim of upholding their health instead of undergoing medical treatment. Southeast Asia is a place rich with a sustainable environment, natural ingredients with great health properties, and an emphasis on traditional medical practices. These are a few of the factors that have facilitated a competitive advantage for Asia as a popular destination for wellness tourism. The world's top wellness tourism destinations are Singapore, Malaysia, Indonesia (Bali), Thailand, and the Philippines.

Digital Transformation in Aviation in Southeast Asia

In the last two decades, advances in technology have played a vital role in all sectors of industry. This advancement in technology in industry is called IR. Technological revolution is a phase in which one or more technologies are replaced by newer technology in a very short period of time. This accelerated technological progress has promotes new innovations which triggered by the rapid application and distribution due to abrupt change in society. Each cycle of revolution increases the industry's productivity and efficiency. An illustration is shown in Figure 1.2.

IR, also known as the First Industrial Revolution (1IR), was the conversion to new manufacturing processes in Europe and the United States. This occurred in the mid-18th century to the mid-19th century, starting in 1784. It was a shift to new manufacturing processes that relied on steam and water. During that time, the first mechanical loom was introduced. 1IR was tremendously advantageous in terms of producing a greater variety of merchandise and improving people's living standards. Coal and steam were fuel for the machinery and made the idea of manufacturing using machinery expand expeditiously. Machines allowed faster and easier production processes, in addition to the growth of a wider range of innovations in technology.

The Second Industrial Revolution (2IR), often referred to as the Technological Revolution, began in 1870 in Britain, Germany, and the United States of America. It was a phase of rapid standardization as well as industrialization from the late 19th century into the early 20th century. With the help of electrical energy, it introduced a division of labour and mass production. During the 2IR, new technological systems were established that allowed for increased output and more sophisticated machines, such as the first assembly line in a Cincinnati slaughterhouse.

The Third Industrial Revolution (3IR) started in the second half of the 20th century (beginning in 1969) and brought forward the expansion of computers,

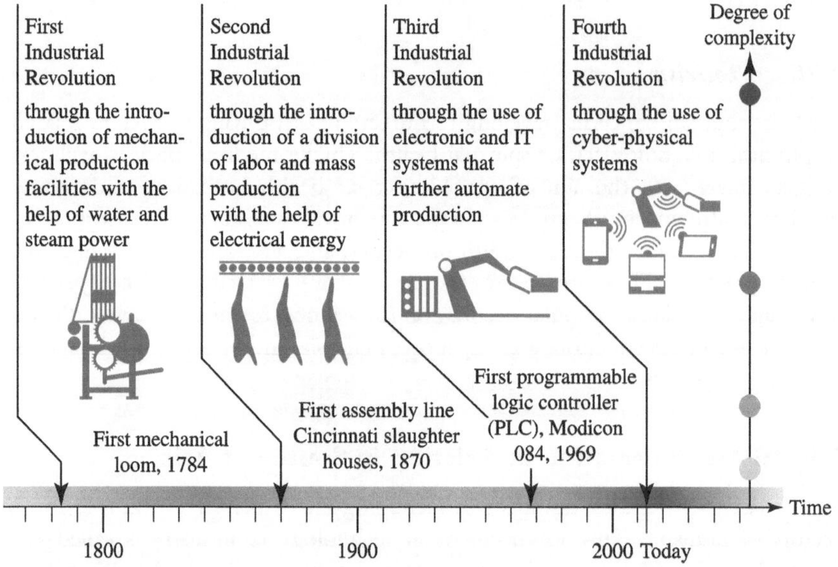

Figure 1.2 The cycle of industrial revolution (source: Supply Chain Game Changer™, 2021).

electronics, and telecommunications. Through these new technologies, 3IR has opened the doors to innovations a majority of which are now available to us to enjoy. Electronics, computers, and information technology (IT) are used to increase industrial automation. Modicon 084, a programmable logic controller (PLC), was one of the automated systems that were introduced during 3IR.

The Fourth Industrial Revolution (4IR) began in the early 2000s and continues to this day. The term Fourth Industrial Revolution was first introduced by Klaus Schwab, the executive chairman of the World Economic Forum (Schwab, 2015). According to him, 4IR includes (but is not limited to) technologies that combine hardware, software, and biology (cyber-physical systems), and advances in communication and connectivity. This is also the era of breakthroughs in emerging technologies such as fifth-generation (5G) wireless technologies, robotics, AI, biotechnology, nanotechnology, quantum computing, IoT, industrial internet of things (IIoT), 3D printing, and autonomous vehicles (Schwab, 2015). 4IR is also the era of smart devices, smart manufacturing facilities, and smart storage systems that are interconnected, activate actions, and control each another without the human interface. IIoT today has made this information exchange possible.

Aviation or airline companies, in addition to catering to their customers' needs and upholding the principles of moment marketing, should also consider real-time influences on their offerings. A platform that will allow the streamlining of every travel micro-moment, from inspiration to post-trip, will be able to help airlines to increase their revenues, build customer loyalty, and improve customer retention. Through big data analytics and machine learning technologies, airline companies can collect passengers' data (e.g. preferences, customer value, inclination to buy a service) and offer personalized offerings to them. This will promote more purchases and retain customers' loyalty to the company (Rose, 2017).

In the most advanced country in Southeast Asia, Singapore Airlines (SIA) together with the National University of Singapore (NUS) have launched a new digital aviation corporate laboratory. This lab will co-create innovative technologies and solutions that will accelerate the digital transformation of Singapore's aviation sector, and help redefine the air travel experience for passengers. Singapore hopes to be at the forefront of digital innovation in the airline industry and create more innovative solutions that can enhance the customer experience and travel journey, optimize revenue generation, and increase operational efficiency using digital technologies (Megha, 2022).

Digital Transformation in Tourism and Hospitality in Southeast Asia

Information technology which started to emerge during the 3IR has played a vital role in the tourism and hospitality industry. This technology, which has brought forth electronics, telecommunications, and computers, has assisted in reducing costs, boosting operational efficiency, and increasing the services level and customer experience. Both tourists and companies can benefit from the better-quality interaction, booking, and customer service systems. This technology also has assisted tourism industries by replacing expensive manpower with technological power. As a result, not only are labour costs reduced, but also customer service issues can be avoided. Subsequently, travel companies are adopting these various technologies in order to improve their operational efficiencies and to meet customers' expectations. Travel companies are using various technologies to engage with customers and assist them at every step of their trip.

Elena Mogoş, Associate Analyst for Digital Travel and Tourism at GlobalData, mentioned that

> Travel still revolves around meaningful human interactions. However, to be successful, companies need to create a balance between the technological innovations and the human element. Improving the customer experience must always be on the back of travel and tourism companies' minds when investing in technology.
>
> (GlobalData, 2018)

Internet Marketing

Internet marketing is an approach used to market products and services via online or other digital channels. These may consist of a variety of online platforms, digital marketing tools, and content delivery systems, such as website content managers, website designers, and mailing list email marketing and social media (Ward, 2020). As in other industries, the internet has a strong impact on the tourism and hospitality industry. The very first experience of a customer with a business is by visiting their website. This includes looking at the photos provided and previous guests' reviews. It is very important for a business to successfully utilize their online advertising, blogs, social media, and online purchasing or booking to provide expediency to customers as a competitive advantage.

Computer Systems

Computer systems are another technology which impacts the tourism industry greatly. Computer systems allow communication not only between tourism service providers and customers but also between hotels in multiple locations of a larger hotel chain. This computer system allows standardization and uniformity between staff in the hotel chain, making it effortless to retrieve information, giving a better experience to the guests. Information regarding reservations, requests from guests, and housekeeping information can be retrieved from the system.

Mobile Communication

The internet has led to so many new innovations, and among them is the way telecommunication is done. Mobile smartphones and tablets have superseded desktop computers, making them almost nonexistent. With mobile devices, the tourism industry has become more popular because travellers can bring their mobile devices on their trip. Travellers will be able to plan their trip, identify tourist sites, take photos, and get information on delays or changes in their reservations, get better trip deals, and find or go to places using GPS tracking apps.

Wi-Fi Connectivity

Wi-Fi or wireless networks enable travellers to stay connected throughout their journey. They can always be connected, either to find ideas for trip destinations, find options on places to visit or have a meal, find directions to tourist spots, or share their moments and experiences through their social platforms. Therefore, travel companies that invest in network services will be able to offer a more seamless and highly personalized feel to customers. This can enhance companies' operational efficiency and customers' immediate decision making, and strengthens the safety and security system (CCTV) or cybersecurity.

Artificial Intelligence

Artificial intelligence or its acronym AI signifies machines that have the capability of simulating human intelligence. These machines are coded to reason like humans and mimic their actions. The term AI may also be applied to any machine that demonstrates characteristics linked with a human mind and the way it thinks, for instance, learning and problem-solving. There are quite a number of advancing technologies and new innovations in the tourism sector resulting from AI. The three major categories of AI that help the travel and tourism industry can be classified into machine learning systems, chatbots or travelbots, and robots. Machine learning systems are an application of AI that provide machines with the capability to automatically learn and assimilate from travellers' experience without being programmed explicitly. With this, the system will be able to predict travellers' demands and expectations. Thus, a much more personalized and specialized service can be offered to them.

Chatbots or travelbots are an AI system that helps tourism companies maintain their good relations with customers. It is more like a machine or system that handles customer service. It is used in many major chat apps such as Telegram, WhatsApp, Slack, Facebook Messenger, etc.

Another form of AI which has impacted travel and tourism industry is robots. Among the human jobs that can and have started to be replaced by robots are hotel staff (reception) as seen in Henn-na Hotel, Nagasaki, Japan; Concierge Robot, a collaboration with IBM called Connie serve at Hilton Hotel worldwide providing information required by hotel guests; an autonomous robotic suitcase called Travelmate; a hotel and airport staff humanlike robot assistant called Shahira; and security robots at airports detecting concealed weapons and other hidden items that are prohibited on flights; to mention a few.

With AI, operations which normally require human intervention and a lot of time to learn new skills can be automated. Consequently, AI speeds up processes and at the same time improves quality and performance and decreases operation costs.

Augmented and Virtual Reality (AR and VR)

In recent years, AR or VR popularity has increased among tourism companies. These technologies are being used either to enhance customers' experiences or marketing. As an example, airlines have begun to use VR technology to demonstrate to travellers what the cabins look like in advance, in order to increase ticket or supplementary services sales.

Internet of Things (IoT)

Internet of things technically refers to every single thing that is connected to the internet, but typically, the term is used more for everyday physical devices, appliances, and other "things" that are connected to the internet to give them the capability to send and receive data. IoT has a great deal of potential in the future of the travel and tourism industry, for example, detecting lost luggage using IoT provided by Lufthansa Airlines. This has really helped reduce passengers' anxiety and stress. Passengers are able to track their luggage via a link found on their mobile boarding pass in the app called Lufthansa.

Voice Technology

Voice technology or voice command systems is a technology that allows devices such as smartphones, computers, smart TVs, and smart home hubs to comprehend human voice commands and respond to them in the correct manner. This technology has started to change the spectrum of the travelling and tourism industry, where more travellers switch from typed-in searches on their mobile devices to using voice commands. Progressively, hotels have started implementing these voice-activated devices. Among those hotels are Westin Buffalo Austin of Marriott International and Kimpton Alexis Hotel.

Verification and Wearable Devices

Other than the technology mentioned previously, tourism companies are progressively adopting technology that is not only able to verify tickets but also provides customers a more tailored and integrated experience. Disney is a company that deploys this technology. They introduced a radio frequency identification (RFID)-equipped wristband called MagicBand which is wearable and customizable (Ngai, 2007). It allows visitors to link to the theme park infrastructure. This will help to reduce guest waiting times, plan their activities, and track their locations.

Existing Possibilities

As mentioned, there are many benefits the travel and tourism industry is able to receive in using technology. The benefits can be enjoyed by both customers and businesses. This technology has improved the communication, reservations, and requests between guests and the company. The application of this technology has aided the tourism and hospitality industry by replacing the costly, high maintenance, perhaps high-error human with technology and systems. Technology has also improved operation efficiency and provides 24 hours' services anytime and anywhere the internet is available. This surely helps businesses to reduce their labour costs and avoid customer service issues. As a result sales and profit increase.

Challenges

It is impossible to overstate just how disruptive COVID-19 has been to the travel industry. The steep decline in tourism and business travel, along with the challenges of maintaining flight crew and passenger safety, are only a few of the problems. The aviation industry has had to adapt to changes and new regulations, often without prior notice.

COVID-19 has created a new norm where people conduct their business or personal activities virtually via online meeting tools or video calls instead of face to face in a meeting room or hall. Businesses have started to become aware of the benefits of remote working and hybrid workforces; these practices are likely to persist, at least in some form, even after COVID-19 has ended. This has contributed majorly to the drop in business travel which is unlikely to rebound anytime soon.

However, the aviation industry continues to respond to its many challenges, despite what seems like a nonstop flood of bad news; if not for technological innovations, it would have taken much longer for the industry to recover and bounce back. In both the present and the future, the aviation industry is well-positioned to meet these challenges.

Future Potential

Aside from the massive disruptions caused by COVID-19, the air and space industries are undergoing major transformations. Business travel will change, supply chain problems will be addressed, and unmanned aerial vehicles (UAVs) will become more popular. The first autonomous flight of a Black Hawk helicopter was launched in February 2022. The increase in satellite coverage may make detection and avoidance more accurate. And as advanced control systems become more advanced, unmanned flight becomes safer.

In another positive light, there have been some promising technological advances, including improvements in "touchless" infrastructure, including passport scanners, face-scanning technology, and more efficient security scanners. In addition to decreasing waiting times in terminals, these technologies could assist the travel industry in its recovery after the pandemic.

Conclusion

The significant contribution of this chapter is to provide an overview of the current market of the aviation industry as well as the tourism and hospitality industry in

Southeast Asia. Southeast Asia is very much associated with tourism, specifically with its unique culture, custom, climate, and geographical characteristics. This has made Southeast Asia one of the best holiday or travel destinations in the world. The number of travellers coming to Southeast Asia increased significantly over the years prior to the COVID-19 pandemic. A majority of the countries have expanded their air terminals as airline companies started to buy more aircrafts to cater to the air passenger traffic coming in and going out of the country. However, although aviation seems to be very lucrative, the COVID-19 pandemic has disrupted and impacted both aviation and the tourism and hospitality industry severely. To help the industries recover after the pandemic, aviation companies/ airlines as well as tourism and hospitality companies have to embrace the adoption of new technology. In the 4IR, new trends in technology innovation and its application have started to play a tremendous part in revolutionizing the aviation and tourism and hospitality industries.

References

Aviation Benefits Beyond Borders (2022). *Tourism Enabler.* Retrieved from: https://aviationbenefits.org/ (accessed: the 31st March 2022).

GlobalData (2018). *Top 6 Technology Trends to Watch Out for in the Travel and Tourism Industry in 2018.* Retrieved from: https://www.globaldata.com/top-6-technology-trends-watch-travel-tourism-industry-2018/ (accessed: the 10th November 2020).

Hall, C. M. and Page, S. (Eds.) (2000). *Tourism in South and Southeast Asia.* London: Routledge.

King, V. T. (2012). Editorial: Culture, Heritage and Tourism in Southeast Asia. *Pertanika Journal of Social Sciences & Humanities*, 20(1), pp. 1–7.

Megha, P. (2022). Singapore Airlines Launches New Aviation Lab to Boost Digital Transformation. *Travel Daily Media.* Retrieved from: https://www.traveldailymedia.com/singapore-airlines-launches-new-aviation-lab-to-boost-digital-transformation/ (accessed: the 1st April 2022).

Mordor Intelligent (2022). *Southeast Asia Aviation Market: Growth, Trends, Covid-19 Impact, and Forecasts (2022–2027).* Retrieved from: https://www.mordorintelligence.com/industry-reports/southeast-asia-aviation-market (accessed: the 1st April 2022).

Ngai, E. W. T. (2007). Mobile commerce integrated with RFID technology in a container depot. *Decision Support Systems*, 43, pp. 62–76.

Nortajuddin, A. (2020). ASEAN Focusing on Domestic Tourism. *The Asian Post.* Retrieved from: https://theaseanpost.com/article/asean-focusing-domestic-tourism (accessed: the 10th November, 2020).

Ringgins, N. (2020). 25 of the Top Business Travel Destinations. *Small Business Operations.* Retrieved from: https://smallbiztrends.com/2016/06/business-travel-destinations.html (accessed: the 15th November 2020).

Rose, D. (2017). Embracing Airline Digital Transformation: A Spotlight on What Travelers Value. *Amadeus.com.* Retrieved from: https://amadeus.com/documents/en/airlines/research-report/embracing-airline-digital-transformation.pdf (accessed: the 1st April 2022), pp. 20–22.

Schwab, K. (2015). The Fourth Industrial Revolution: What it Means and How to Respond. *Foreign Affairs*. Retrieved from: www.foreignaffairs.com/articles/2015-12-12/fourth-industrial-revolution (accessed: the 24th August 2021).

Singapore Tourism Board (n.d.). *Annual Report 12/13*. Retrieved from: http://www.stb.gov.sg/htm/STBAnnualReport/STB_AR12-13/pdf/stb_ar_2013.pdf (accessed: the 10th November 2020).

Southeast Asian Regional Centre for Graduate Study and Research in Agriculture (SEARCA) (2019). *Philippines Recognized as a Top Agritourism Destination*. Retrieved from: https://www.searca.org/press/philippines-recognized-as-a-top-agritourism-destination (accessed: the 7th November 2020).

Standing Committee for Economic and Commercial Cooperation of the Organization of Islamic Cooperation (COMCEC) (2016). *Muslim Friendly Tourism: Understanding the Demand and Supply Sides in the OIC Member Countries*. Retrieved from https: //sbb.gov.tr/wp-content/uploads/2018/11/Muslim_Friendly_Tourism_Understanding_the_Demand_and_Supply_Sides_in_the_OIC_Member_Countries%E2%80%8B.pdf. (accessed: the 22nd November 2020), pp. 1–2.

Supply Chain Game Changer™ (2021). *The Industrial Revolution from Industry 1.0 to 5.0*. Retrieved from: https://supplychaingamechanger.com/the-industrial-revolution-from-industry-1-0-to-industry-5-0/ (accessed: the 11th November 2020).

United Nations World Tourism Organization (UNWTO) (2017). *Affiliate Members Report, Volume Fifteen: Second Global Report on LGBT Tourism*. Retrieved from https://www.e-unwto.org/doi/epdf/10.18111/9789284418619 (accessed: the 20th November 2020), pp. 62–63.

US Department of Economic and Social Affairs (2010). *International Recommendations for Tourism Statistics 2008*. Retrieved from: https://unstats.un.org/unsd/publication/Seriesm/SeriesM_83rev1e.pdf#page=26 (accessed: the 22nd November 2020), p. 15.

Ward, S. (2020). What is Internet Marketing? Definition & Examples of Internet Marketing. *The Balance Small Business*. Retrieved from: https://www.thebalancesmb.com/internet-marketing-2948348 (accessed: the 10th November 2020).

Winter, T., Teo, P. and Chang, T. C. (Eds.) (2009). *Asia on Tour. Exploring the Rise of Asian Tourism*. Abingdon: Routledge.

World Travel Tourism Council (WTTC) (2020). *Economic Impact Report: The Importance of Travel & Tourism in 2019*. Retrieved from: https://wttc.org/Research/Economic-Impact (accessed: the 19th November 2020).

2 Digital Transformation and the ASEAN Influences, Issues, and Challenges

Au Yong Hui Nee

Introduction

Based on United Nations World Tourism Organization (UNWTO) data, Southeast Asia received tourism receipts worth US$130.7 billion in 2017 (Yeo, 2019). International tourist arrivals in the region of Asia and the Pacific accounted for one-quarter of global arrivals, and this is expected to expand with more middle-income earners and the availability of budget airlines (Singapore Business Review, 2018). In 2018, the Association of Southeast Asian Nations (ASEAN) region logged more than 135 million international visitors, where the top countries by international tourist arrivals are Thailand, Malaysia, Singapore, Indonesia, and Vietnam.

The global travel and tourism industry is affected by global trends on top of economic, political, technological, and cultural elements. Digital technology, cloud technology, information and communication technology (ICT), and artificial intelligence (AI) can be useful in many economic sectors from the medicine sector to schools to the manufacturing and tourism sectors (Astawa et al., 2019).

In tourism, technology has connected destinations and people, reduced the costs of marketing, and reduced the prices of travel. The choices in accommodation have also increased for travellers, reducing the cost of stay. Travellers can benefit from Airbnb arrangements. Travellers can collect all the information they need for tour planning. Travelling is safer, and travellers can remain connected. Hospitality is providing services with a personal touch with the use of advanced technology, data, and analytics (Pratap, 2021).

Research shows fewer innovations in tourism because of the insufficiency of knowledge and roles in sustainable tourism, inadequate cooperation, and scarcity of incentives. The geographical region of origin influences health travellers' behaviour intention (Na et al., 2017). Furthermore, travellers have led the industry to cater to a client base searching for local experiences and customized services (Salazar, 2018).

Digital Transformation

The rapid diffusion of broadband, in particular mobile phones, is enabling more people to connect to communication networks in Southeast Asia. The high level of mobile phone subscriptions in almost all ASEAN countries has made

DOI: 10.4324/9781003315179-4

communication and connectivity easy, and this priority area indicates an opportunity to improve interconnectivity through digitalization (Suriyankietkaew and Nimsai, 2021).

ASEAN countries have implemented several digitalization initiatives. Governments can link up payment vendors with businesses and get more small traders to link to online platforms to spur trade and economic activity (Vandenberg, 2020). For instance, the Philippines' digital strategy focuses on improving internet connectivity by reducing the geographical digital divide and ensuring affordability. The Philippines is also focusing on upgrading government ICT infrastructure to tolerate integrated government operations (Menon, 2016). The Thailand government-initiated money transfer platform under the national e-payment scheme, PromptPay, was launched in 2017; its user registrations had increased to 55.6 million by the end of September 2020, growing 11.4 per cent from 49.9 million in 2019, according to the Bank of Thailand (Koetsier, 2020).

A World Bank report on Malaysia's digital economy (Record et al., 2018) suggests the government changing its role from offering entrepreneurship support programmes to tackling structural reforms, while preventing crowding out effects, channelling government funds into ineffective programmes, or putting burdensome rules in position. In contrast, based on Alpha JWC Ventures (2019), Indonesia is the biggest contributor to the development of the digital industry in Southeast Asia.

Due to rigorous COVID-19 pandemic public health policies, the effort to attain the United Nations (UN) sustainable development goals (SDGs) or Agenda 2030 for sustainable development in Southeast Asia has been interrupted (Suriyankietkaew and Nimsai, 2021). However, the pandemic has not deterred but accelerated the digital transformation. The Asian Development Bank (ADB) has adjusted its assistance to get more small- and medium-sized enterprises (SMEs) to employ digital payments to increase financial inclusion (Sawada and Sumulong, 2021).

The ASEAN

The ASEAN is a region with a relatively young population. Fifty per cent of ASEAN's population is under 30 years of age, and 90 per cent of these people have access to the internet; however, according to Menon (2016), none of the ASEAN member countries, except for Singapore, Malaysia, and the Philippines, have a comprehensive digital strategy in place. Malaysia has proclaimed a Digital Malaysia 354 (DM354) Roadmap that represents three ICT focus areas, five subsectors, and four Digital Malaysia communities. The DM354 Roadmap concentrates on big data analytics and education, and other sectors (Asohan, 2012).

During the COVID-19 pandemic, the endeavour of achieving SDGs continues. ASEAN's regional digital agenda has a strong focus on e-commerce rather than the digitalization of industries. ASEAN has accepted two UN Commission on International Trade Law (UNCITRAL) instruments, the Model Law on

Electronic Commerce (MLEC) and the Convention on Electronic Contracting (Beschorner, 2021). Governmental bodies and the private sector of ASEAN have started to create a new initiative, the "National Digital Trade Platform (NDTP)," in the CLMV countries (i.e., Cambodia, Laos, Myanmar, and Vietnam) to help the digitalization of intra-regional trade and e-commerce (Suriyankietkaew and Nimsai, 2021).

Menon (2016) suggests that ASEAN's regulatory frameworks need to be overhauled to enable more innovation in financial transactions, taking advantage of the extensive use of smartphones. There are three sets of regulations that are significant for adopting mobile financial services: first, regulations to develop mobile financial services; second, regulations relating to cybersecurity, privacy, data protection, and e-signatures; and third, digital payment regulations that are synchronized in ASEAN to accelerate cross-border trade and remittances.

ASEAN member states progress in various stages. In the last decade, Malaysia, Singapore, and the Philippines embraced comprehensive data protection laws. Thailand's personal data protection law (Beschorner, 2021) came into effect in 2021. Digital ID is connected with development outcomes, e.g., comprehensive social safety nets, rationalized public administration, financial inclusion, and the empowerment of women. Brunei, Indonesia, Malaysia, Singapore, and Thailand have fully digitalized their foundational ID system and have created a related public key infrastructure (Beschorner, 2021).

The ASEAN Influences on Digital Transformation

The ASEAN region is mostly connected to the internet via mobile devices. It is estimated that 78 per cent of the population in Thailand will have access to the internet by the end of 2021 (Statista.com, 2022). In 2019, the rates of the population who had access to the internet were 70 per cent for Vietnam, 48 per cent for Indonesia, and 31 per cent for Myanmar (World Bank, 2020). It was reported that the people in Southeast Asia devote more time to accessing internet information with their mobile devices than the global average. Southeast Asia leads the world in mobile internet use. Specifically, the Thais use mobile internet more than five hours a day, compared with over four hours in Indonesia, the Philippines, and Malaysia, and three hours globally, based on Google (Nguyen, 2019).

To enable wider participation in the digital economy, ASEAN must reduce its reliance on physical cash (Menon, 2016). "Inclusive" digital transformation can be achieved by helping micro-, small-, and medium-sized enterprises (MSMEs) access supply chains and enhancing consumer welfare through transactions on digital platforms (Sawada and Sumulong, 2021). A high mobile phone penetration rate also promotes digital transformation. According to Naishtein (2018), digital payment services are fast becoming the norm in Singapore, as three in five merchants accept e-payments, and there are plans to make the country check-free by 2025. In the case of the Philippines, digital finance is escalating competition in the financial sector and directing wider access to funds (Sawada and Sumulong,

2021). In Indonesia, digital platform businesses have triumphed in achieving excellent investment value both as decacorns and unicorns, including Gojek, Grab, Tokopedia, Bukalapak, Halodoc, and Ruangguru (Technologue.id, 2020).

The ASEAN Issues in Digital Transformation

Digital transformation takes time in the ASEAN. Even though the Brunei government declared the possibility of venturing into augmented reality (AR) and virtual reality (VR) at the ASEAN Tourism Forum 2020, there is no plan to follow in creating virtual tourism (Scoop, 2020). Based on Sathirathai and Wan (2018), e-commerce penetration in ASEAN averaged about 2 per cent of the population in 2017, which is considerably less than China and the United States, at about 23 per cent and 8 per cent, respectively. The inadequate use of digital technology may reveal the digital divide between large and smaller firms, as generally SMEs lag in technology adoption in Southeast Asia (Organisation for Economic Co-operation and Development, 2017).

The ASEAN Challenges in Digital Transformation

The challenges for Southeast Asia from the accelerated pace of digitalization caused by the COVID-19 pandemic are remarkable and will ultimately affect the government more than the private sector, though they will disrupt both significantly (Mondejar et al., 2021).

Nevertheless, there are still some people even in Singapore who are still reluctant to adopt cashless payments (Naishtein, 2018). Factors limiting the use of digital technology in ASEAN countries include the limited availability of affordable broadband internet (Menon, 2016; Organisation for Economic Co-operation and Development, 2017; Purbasari et al., 2021), the lack of necessary digital skills (Organisation for Economic Co-operation and Development, 2017; Purbasari et al., 2021), and network security (Purbasari et al., 2021).

Investment in Tourism

For the tourism sector to develop and foster, it requires capital, infrastructure, knowledge, and marketing access and distribution channels (Banerjee et al., 2015; Hof and Blázquez-Salom, 2015; Stauvermann and Kumar, 2017). In the literature on tourism, technology includes not merely the "hardware" but the "software" of the tourism and hospitality industry. Rather than skills, the United Nations Conference on Trade and Development (UNCTAD) (2007) emphasizes the advertising power of international brands; destination marketing organizations can profit from the loyalty schemes, corporate networks, and client bases of international firms. UNCTAD (2007) further suggests that these international firms have the capacity to enhance their international recognition and boost their image. Furthermore, international firms also promote standards with more developed systems and quality control (Nunkoo and Seetanah, 2018).

Tourism is a major foreign exchange earner in many countries. The higher the economic growth, the more receipts in tourism (Au Yong, 2018). Bhrammanachote (2018) concentrated on an eclectic paradigm, or the OLI Model, which stands for "Ownership, Location, Internationalization," originated by John H. Dunning. The chapter adds value to the literature by showing that the local elements matter for international investment.

Nonthapot (2017) further suggests a bi-directional long-run causality between principal investment in the tourism sector and arrivals of international tourists. There is also a strong correlation between the foreign tourists and the investments that took place (Snyman, 2007). The effects of foreign direct investment on the tourism sector were detected in tourism-related sectors such as tourist arrivals, hotels, and tour agencies (Bounvilay, 2015). Results obtained by Falk (2014) show that hotel foreign direct investment (FDI) decreases with business laws, tax rates, and salaries, and improves with market share. Consequently, the roles of overseas development assistance (ODA) and FDI are also essential for the digital economy, referring to the experience in South Korea, Taiwan, and the Philippines (Astawa et al., 2019).

The results of Perić and Radić (2016) present how a foreign-owned hotel in a developing country has a significant impact on the host economy. Foreign direct investment was found to be focused on upscale hotels which were inclined to hire more staff, give higher salaries, and arrange more training. They have a lower percentage of female workers, but more high-skilled jobs, and more expatriate managers. On the other hand, some international owners were active in best practice community-linked enterprises driven by environmental and social morals (Davidson and Sahli, 2015). Tourism is appreciated as a means of earning foreign exchanges, creating large numbers of jobs, encouraging economic diversification and a service economy, reviving urban zones and cultural activities, and opening up remote rural zones (United Nations Conference on Trade and Development, 2007). The results from Radić(2018) show that terrorism has no significant effect on tourism FDI inflow while international tourist arrivals shape future FDI. The tourism industry has to create more jobs than it destroys, complement local industries, and maintain the cultural environments' integrity (Haley and Haley, 1997). Widiatedja (2019) argues that trade and investment agreements may include provisions to improve environmental impacts due to tourism. As a result, the basic elements used in recognizing the uniqueness of the green hospitality application are based on the ASEAN Green Hotel Standard and EarthCheck (Astawa et al., 2019).

Reported by the *Bangkok Post* (2013), air travel growth along with improved infrastructure has enhanced the links between Southeast Asia and the rest of Asia. Cambodia is at the peak of economic growth with increasing visitor arrivals and foreign direct investment from South Korea, Vietnam, and China.

The main emerging tourism destinations in Southeast Asia are Vietnam and Cambodia. Vietnam's flourishing tourism sector and booming economy have reinforced the hotel and resort market, with foreign investors showing interest in the country (Thom, 2019). FDI sites such as Phu Quoc Island are eligible to apply for the highest investment preference policies. In recent years, the hotel sector in Vietnam has caught the attention of foreign investors, mostly from Japan, South Korea, Hong Kong, and Singapore (Anh, 2022). Other factors, such as visa exemptions, new direct air routes, and better marketing efforts have also enhanced its attractiveness to leisure travellers. While foreign hospitality investors are seeking higher returns, the domestic investors are interested in greenfield development (Thom, 2019). Reported by *Vietnam Economics News*, international hotel chains such as AccorHotels, Hilton Hotels, InterContinental Hotels, Marriot International, and Starwood Hotels are providing capital into Vietnam, while other industry players such as Wyndham Hotels, Holiday Inn, and Pan Pacific Hotels are projected to enter too (*Vietnam Economic News*, 2016).

Ministry of Tourism data shows the importance of tourism in the economy of Cambodia due to ticket sales and tourists' expenditure. Cambodia has attracted substantial foreign direct investments into the tourism, hospitality, and food and beverage (F&B) sectors. This has pushed its economy to become more liberal to meet the greater demand for products and services, leading to economic expansion and higher revenue for the country. Regulatory developments and incentives are implemented to allow the economy to thrive at a fast sustainable pace. Upmarket hotels are eligible for import duties exemption, and profit tax exemption. Tourism also contributes to human resource development through job opportunities and skill-based training (B2B Cambodia, 2016). According to the Cambodia Association of Travel Agents, Cambodia wants a hospitality industry that follows hygiene and quality standards. For the Cambodian National Tourism Alliance, even though the industry is mostly concentrated in Siem Reap, Phnom Penh, and Sihanoukville, other destinations can also be considered. Furthermore, vital tourism destinations in the country need sufficient hospitality institutions to meet demand during the busiest periods (Chan, 2018). Based on Viet An Law Firm (2017), the next boom will be in Sihanoukville, or "Kampong Som," a coastal city in Cambodia. Phnom Penh and Sihanoukville are the primary real estate development beneficiaries. Mid-tier cities such as Poipet and Battambang are also expected to see rises in property prices, contributed by urbanization and growth in Asia (Tang and Wong, 2011). The results show that FDI does not merely promote exports but also promote vertical linkages in Cambodia that could create positive externalities. There is a need to increase the human resources in tourism professionalism, from reviewing policies to establishing or running public training programmes together. Therefore, Nam (2018) recommends strategic plans for local human resource development in foreign businesses.

E-tourism refers to ICT applications in the tourism industry. Industry players such as tour operators, tour agencies, hotel establishments, car rental companies, cruise companies, and others can gain from ICT benefits in marketing, sales, operations, human resource management, and purchasing (Hojeghan and Esfangareh, 2011). Velikova (2019) shows that tourist spots concentrating on innovative supply to their clients utilize up-to-date technologies directly involved in expanding the tourism services supply. Consumer value is improved by offering place, promotion, and product innovations. A content analysis by Law, Leung,

Au, and "Andy" Lee (2013) learnt that more than half of the articles reviewed had been concentrated on the hospitality sector in general and that a considerable number of IT-related articles had been published since the 1980s. In Malaysia, the Bank Negara Malaysia (BNM) has published its Financial Sector Blueprint 2011–2020. The BNM aims to raise e-payment transactions per capita to 200 transactions (Bank Negara Malaysia, n.d.).

Technology Applications in the Tourism and Hospitality Industry

Technology's intrinsic characteristics are the key element in the adoption of information technology (IT). De La Santa and Agatep (2005) affirmed that firm ownership and size positively affect technology adoption and recognized the power of social and culture that shape its adoption. Jokonya (2019) and Entre (2018) revealed that ICT plays a major role in the hospitality and tourism industry, mostly in general forms such as email, social media, smartphones, laptop and desktop computer systems, internet marketing, Wi-Fi, and mobile technology.

Hotels can utilize technology for an improved customer experience through mobile phones, tablets, self-check-in, live chat, and door opening technology (Pathak, 2015). According to Lubbe and Douglas (2019), the new technologies that are redesigning markets include the sharing economy, the internet of things, autonomous vehicles, AI, blockchain technology, and big data analytics. The sharing economy and the experience economy powerfully impact the initiation of new business models that integrate business processes, the power to adjust to new conditions quickly, and digital interfaces (Pindţo and Barjaktarović, 2018). These digital transformation variables have additional impacts on the revenue growth rates because the trends are based in large part on the capabilities to communicate and create mass marketing in tourism (Ribeiro and Florentino, 2016).

Digitalization is desired for tourism establishments to meet the ever-growing expectations of clients (Happ and Ivancsó-Horváth, 2018). The digital themes and initiatives of the future are seamless journeys tailored to their habits and preferences, empowering the travel ecosystem, digital enterprise, and safety and security (World Economic Forum, 2017). Härting, Reichstein, Härtle, and Stiefnomics (2017) show six main drivers of digitalization, which are sales growth, booking, the sharing economy, personalized offers, social media, and customer reviews. Consequently, more hotels are aiming at online promotion. Hotels have to develop their online sales features, e.g., direct booking and online tour agencies (Sabinezs, 2018).

Reported by the Malaysian Reserve, the Deputy Secretary-General of Tourism said the country needs to push digitalization for transactions from about 30 per cent presently. Digital-based technology is required for the change towards an e-payment society like in China, Japan, and South Korea. In other words, these businesses need to connect to digital platforms used by foreign tourists. Having improved online visibility and accepting common payment methods, these businesses can increase their occupancy rates and revenues which translates to more growth and jobs in destination countries. Airbnb is also an innovation to which conventional hospitality will have to react as its effect has led to a demand for laws (Oskam and Boswijk, 2016).

The most important functionalities of near field communication that are predictable and some example applications and trials in the tourism setting were in the airline industry, the hotel and catering industry, and destinations (Egger, 2013). Findings from Floricic and Pavia (2017) verify the enthusiasm of youths to lodge in an innovation test hotel, to evaluate innovations, and to spend the market rate for this special experience of playing a role of the process of innovation.

Based on Stubbs (2018), the five ways technology is changing consumer expectations in hospitality are that things are done quicker, customers' needs are anticipated, and there is shareability, two-way relationships, and accessibility of services. According to Lyons (2017), a PwC report presented tourism and hospitality as the fifth industry with the highest investments in sensors in 2014. Inferred by the report, this investment is vital for customer experiences. Technology improves efficiencies via the automation of check-ins and check-outs, and keyless entry; creating a more personal experience via technology saves energy, as do robotics in hotels; and technology develops experiences via online bookings and demand for data analysts and scientists. In addition, according to Saratchandran (2018), the six ways hotels can use technology are online bookings, online reputation, improving guest experiences, enhanced targeting, streamlining internal processes, and smart solutions.

The latest field of information and communications technology is important in tourism; it offers intelligent agents and the sematic web, while virtual reality is no longer fiction even in the most remote places on the earth (Soava, 2015). Technology-driven tools and platforms are operating as catalysts for rapid digitalization of the travel and hospitality sector (KPMG in India and FICCI, 2018). Thus, the potential of digitalization is in checking recommendations, online check-in and check-out, secure reservation and purchase process, application development, smart cities, connectivity, access to devices, new business models, sector-based focus, and augmented and virtual reality.

Mariani (2020) proposes that tourism and hospitality scholars are gradually alert to and embracing big data (BD) and BD analytics such as AI, and the internet of things (IoT). Businesses apply AI to cut down operating costs and maintenance charges. AI-oriented marketing has already been extensively applied. The tourism and hospitality industry will employ big data and AI to satisfy customers, and to transform the industry (Çeltek and Ilhan, 2020). Digitalization, big data, and cloud computing will make it simpler to cognize customer needs more precisely. With Industry 4.0 (I4.0) applications, reservations, room selection, and F&B can be pre-ordered. Intelligent robots may be able to propose services such as porter service, food and beverage orders, greeting, on-site guidance, payment, promotion, and transfer (Gül et al., 2018).

Choi, Oh, Choi, and Kim (2021) specify that interfacing with robots is a major element in robot-operated hotels, where Japanese appreciate more on emotional responses while non-Japanese on technical features. Consequently, robot-friendliness would be a new competitive advantage (Ivanov and Webster, 2017b). In particular, Revfine (n.d.b) provides examples of hospitality robots, such as the world's first robot-operated hotel, the Hilton robot concierge, a robot suitcase, a robot assistant, a travel agent robot, a chatbot, and security robots for airports.

Bhattacharjee, Seeley, and Seitzman (2017) suggest that the hospitality sector can use artificial intelligence to create opportunities from data through predictive analytics, natural-language processing, virtual assistants, robotics, and smarter pricing management, among others. According to Revfine (n.d.a), four out of ten hospitality trends are related to technology applications (i.e., smart hotels, robots, virtual reality, and augmented reality). A glimpse into the future trends shows development in artificial intelligence and underlying technologies, advances in artificial intelligence, the regulation of artificial intelligence systems, and talent development (Geisler, 2018). Based on Saulat (2018), the technology trends are digital interactions, facial recognition, machine learning, and social media. Gartner predicts that one in five workers will rely on AI by 2022, for instance, custom voice programmes/voice-assistants, speak-to-order, augmented reality, and biometric authentication. Hospitality will adopt both human interaction and more advanced technologies (Kobres, 2018).

The idea of building intelligent hotels involves the utilization of automatic data processing (machine learning), concierge robots, digital assistance, travel experience enhancers, and voice-activated services. According to Marutitech (2020), AI inputs in the hotel industry include bookings, choices, itineraries, preferences, ratings, and payments. In other words, the industry is increasingly beginning to embrace robots, AI, and service automation technologies (Ivanov, 2019).

Technology Applications in the Southeast Asian Tourism and Hospitality Industry

Social media platforms are closely related to the tourism and hospitality industry. The results from Bui (2021) show over 80 per cent of travellers book a trip online. In addition, travellers in their 30s and 40s want access to more social media feedback and ask for recommendations from friends. The most utilized social media platforms are Instagram, Facebook, and YouTube. These travellers also use TikTok and Pinterest quite regularly, especially those aged 30–40 years. In addition to concentrate on social networking sites such as Facebook, Instagram, and YouTube, TikTok and Pinterest are still considered value investing in the tourism and hospitality industry.

In the tourism and hospitality industry, technologies can play many roles. Since 2017, Vietnam's National Administration of Tourism has partnered with the national telecommunications group in rolling out smart tourism projects (Vietnamplus, 2019). An instance of smart tourism development in Thailand includes equipping airports with advanced technologies and building data platforms that permit traveller monitoring at destinations such as Phuket (Phocuswire, 2019). From the supply side, one of the hotel technology providers is Japan-based TradFit. It provides hotels with a digital concierge applying artificial intelligence, IoT, and voice recognition to convert the hotels and enable them to focus on delivering quality service (Yeo, 2019).

The application of technology in the industry brings benefits to the clients as well as the hotel staff. Technological innovations are such as big data, e-commerce, and e-wallets and internet of things. The *Nikkei Asian Review* reports that Alipay and WeChat Pay are expanding their market to Southeast Asia. Thai businesses are incorporating Alipay and WeChat Pay's systems to provide to travellers from China and integrating them into local transport systems. However, piracy, privacy risks, scams, and security concerns could possibly hurt their businesses (*The ASEAN Post*, 2018).

In particular, Singapore is leading the Southeast Asian countries in adopting technology applications, as the drive for innovation is led by the Singapore Tourism Board (STB). Most of the functions replaced are those related to housekeeping, deliveries, or butlers, such as in Hotel Jen Tanglin Singapore, Park Avenue Rochester Hotel, Yotel Singapore, Sofitel Singapore City Centre, and M Social Singapore (Singapore Business Review, 2018). Another instance of technology application adoption is W Singapore Sentosa Cove Sentosa Island (Site Minder, n.d.). Besides these, there are also self-service technologies that give the chance for hotels to select an application that fits their clients (Ong, 2010).

Digital Transformation Issues and Challenges in the Southeast Asian Tourism and Hospitality Industry

Ivanov and Webster (2017a, 2019) point out the necessity to attain a balance of the cost-benefit ratio in the adoption of technology. While the cost of software is the main consideration, firms consider social media marketing in terms of customer satisfaction and engagement (Benea, 2014), and hotel/hospitality revenue management (RM) software in terms of timesaving, assisting decision-making, and generating more income (Alrawadieh et al., 2020). In the adoption of e-tourism, there are several benefits including flexibility and interactivity, and challenges like connectivity issues and redundancy of intermediaries (Kiprutto et al., 2011); as for service robots in hotels, these improve efficiency and productivity, and create issues in terms of costs, personnel skills, and substantial changes to organizational culture and structure (Xu et al., 2020).

Among the ASEAN countries, most international tourists visit Thailand, Malaysia, Singapore, Indonesia, and Vietnam. However, the tourism industry in Thailand faces new challenges and opportunities in relation to the high-value segment of older tourists, competition from other destinations, the increase in online services, the use of technology that reshapes travel marketing, and emerging business models (e.g., Airbnb) (Siam Commercial Bank Economic Intelligence Centre, 2017). On the other hand, the study of Salleh et al. (2010) recognizes the operating challenges in revenue, human resources, and F&B, while marketing is likely to be the main challenge. In the future, the challenges expected to be confronted are in terms of regulatory challenges, market competition, and personnel professionalization.

Singapore hotels are developing advanced technology measures focusing on invention, efficiency, and talent. Far East Hospitality is the pioneer STB associate transforming its business with advanced technology solutions, targeting reduced manpower and supporting valuable customer experiences, performing more innovative tasks and human interaction (Cher, 2018). The Indonesian hotel industry is facing more challenges as digital disruption continues to overturn conventional business models. Widiatedja and Wairocana (2018) show that the existence of foreign direct investment has been handled through fragmented and unclear laws based on Indonesian laws of investment, tourism, environment, labour, and land. The industry is facing a surplus of hotel rooms and inadequate certification standards. The chairman of the Indonesia Hotels and Restaurants Association (PHRI) said digitalization has played an important role in boosting hotel room reservations. However, these foreign online agencies don't seem to contribute to income taxes, though they impose more fees than traditional tour companies (Al Azhari, 2017).

Tourism in Vietnam has been a contributor to social development and especially to poverty alleviation since 2005. However, some of the tourist numbers and revenue figures have not been met, mainly due to the dramatic decline following the dispute over the South China Sea for the Chinese market and the decline of the rouble for the Russian market. Other barriers to generating more tourist numbers include concerns over pollution in tourist destinations, limited tourism products, second-rate facilities and amenities, and poor customer service (Truong and Le, 2016). In terms of technology trends, electronic commerce is also likely to increase for online services in tourism, entertainment, gaming, and education. Based on the Vietnam E-Commerce Association, the room reservations rate through online travel agents has an annual growth rate of over 30 per cent since 2016. However, 80 per cent of the market share is controlled by international players (e.g., Agoda and Booking) (EU-Vietnam Business Network, 2018). In addition, user-friendly interfaces, simple, fast page load speeds, limited interruptions, secure payment systems, and reduced risks when paying using PayPal are all vital technological challenges to overcome (Bui, 2021). Furthermore, according to Lu (2019), customers acknowledged that employing AI provides always-on service, fast access to service and cleanliness; nevertheless, customers would still desire to interact with workers.

Conclusion

Tourism is important to a country as a contributor to social development and especially to poverty alleviation. The impact of the fourth technology revolution has been enormous, affecting industries such as tourism and hospitality. The industry is applying technological changes in the way it provides customer service to transform the clients' travel experience and reinvent business models. The use of smart technology in the industry brings benefits to the clients as well as employees of the industry. The industry uses smart technology to reduce operating costs, for a better guest experience, and to create opportunities from big data. One of the common usages of the latest technology that brings success is room reservations through online travel agents. In some countries, the rate of adoption of smart technology is much higher than in neighbouring countries, due to the drive for technology innovation led by their tourism promotion agencies. Other countries can emulate these to promote suitable cost-effective smart technology for further development of the industry.

ASEAN is a region with a relatively young population, and many of these young people have access to the internet. The high level of mobile phone penetration has made communication and connectivity easy. ASEAN's regional digital agenda focuses more on e-commerce than the digitalization of industries. In view of the SDGs, "inclusive" digital transformation can be achieved through transactions on digital platforms. Actually, some digital platforms originating in the region have achieved great investment value both as decacorns and unicorns. Therefore, governments are recommended to change their role in tackling structural reforms, as there are still remaining issues such as the digital divide and challenges in terms of the limited availability of affordable broadband, the lack of necessary digital skills, and network security.

References

Al Azhari, M. (2017). *Digital Disruption Slows New Investments in Hospitality Industry: Indonesian Hoteliers.* Jakarta Globe. Retrieved from: https://jakartaglobe.id/context/digital-disruption-slows-new-investments-hospitality-industry-indonesian-hoteliers/ (accessed: the 17th November, 2021).

Alpha JWC Ventures (2019). *Pertumbuhan Ekonomi Digital Indonesia Menjadi yang Paling Pesat di ASEAN.* Retrieved from: https://www.alphajwc.com/en/pertumbuhan-ekonomi-digital-indonesia-menjadi-yang-paling-pesat-di-asean/ (accessed: the 6th December, 2021).

Alrawadieh, Z., Alrawadieh, Z. and Cetin, G. (2020). Digital transformation and revenue management: Evidence from the hotel industry. *Tourism Economics*, 27(2), pp. 328–345.

Anh, N. (2022). Global hotel giants flock to Vietnamese tourism. *Vietnam Investment Review*. Retrieved from: https://vir.com.vn/global-hotel-giants-flock-to-vietnamese-tourism-62515.html (accessed: the 16th November, 2021).

Asohan, A. (2012). Digital Malaysia details out ... finally! *Digital News Asia*. Retrieved from: https://www.digitalnewsasia.com/digital_economy/digital-malaysia-details-out-finally (accessed: the 17th November, 2021).

Astawa, I. K., Budarma, I. K., Widhari, C. I. S., Mudana, I. G., Murni, N. G. N. S. and Suardani, A. A. P. (2019). Green hotel as a competitive product (case study at a 5-star hotel in Bali). In The 3rd Asia Pacific Management Research Conference. Bali: 13th–15th November 2019.

Au Yong, H. N. (2018). Positioning Malaysia in medical tourism: Implication on economic growth in ASEAN integration. *Indonesian Journal of Business and Economics*, 1(1), pp. 49–73.

B2B Cambodia (2016). *Investment Insight: A Perspective on What Growth in Tourism Means to Cambodia.* Retrieved from: https://www.b2b-cambodia.com/articles/investment-insight-a-perspective-on-what-growth-in-tourism-means-to-cambodia/ (accessed: the 17th November, 2021).

Banerjee, O., Cicowiez, M., & Gachot, S. (2015). A quantitative framework for assessing public investment in tourism – an application to Haiti. *Tourism Management*, 51, pp. 157–173.

Bank Negara Malaysia (BNM) (n.d.). *Payment Systems: Driving towards Electronic Payments.* Retrieved from: https://www.bnm.gov.my/driving-towards-electronic-payments (accessed: the 17th November, 2021).

Benea, I. A. (2014). *Influences of Social Media on the Tourism and Hospitality Industry.* Bachelor Thesis. Vienna: Modul University.

Beschorner, N. (2021). The digital economy in Southeast Asia: Emerging policy priorities and opportunities for regional collaboration. In Christopher Findlay and Somkiat Tangkitvanich (Eds.) *New Dimensions of Connectivity in the Asia-Pacific.* Acton: ANU Press, pp. 121–156.

Bhattacharjee, D., Seeley, J. and Seitzman, N. (2017). *Advanced Analytics in Hospitality.* McKinsey Digital. Retrieved from: https://www.mckinsey.com/business-functions/mckinsey-digital/our-insights/advanced-analytics-in-hospitality https://www.b2b-cambodia.com/articles/investment-insight-a-perspective-on-what-growth-in-tourism-means-to-cambodia/ (accessed: the 17th November, 2021).

Bhrammanachote, W. (2018). Determinants of foreign direct investment in Thailand: Chinese investors in Chiang Mai, Thailand. *PSAKU International Journal of Interdisciplinary Research*, 7(2), pp. 160–173.

Bounvilay C. (2015). *The Foreign Direct Investment (FDI) Promotion Policies and Tourism Development: A case study in Luang Prabang Province Lao People's Democratic Republic.* Master of Science Report. Kuala Lumpur: Asia Pacific University (APU).

Bui, H. (2021). *Applying the 4Ps Marketing Mix in the Vietnam Hospitality Industry to Attract European Tourists.* Bachelor's Degree in International Business Thesis. Joensuu: Karelia University of Applied Sciences. Finland.

Çeltek, E. and Ilhan, I. (2020). Big data, artificial intelligence, and their implications in the tourism industry. In E. Çeltek (Eds.), *Handbook of Research on Smart Technology Applications in the Tourism Industry.* Hershey: IGI Global, pp. 115–130.

Chan, S. (2018). Hospitality establishments continue to increase. *Khmer Times.* Retrieved from: https://www.khmertimeskh.com/109870/hospitality-establishments-continue-increase/ (accessed: the 11th November, 2021).

Cher, B. (2018) Hospitality industry faces challenge of integrating technology to pursue growth. *The Edge Singapore.* Retrieved from: https://www.theedgesingapore.com/news/state-city/hospitality-industry-faces-challenge-integrating-technology-pursue-growth (accessed: the 15th November, 2021).

Choi, Y. J., Oh, M., Choi, M. J. and Kim, S. S. (2021). Exploring the influence of culture on tourist experiences with robots in service delivery environment. *Current Issues in Tourism*, 24(5), pp. 717–733.

Davidson, L. and Sahli, M. (2015). Foreign direct investment in tourism, poverty alleviation, and sustainable development: a review of the Gambian hotel sector. *Journal of Sustainable Tourism*, 23(2), pp. 167–187.

De la Santa, E. I. and Agatep, M. V. D. (2005). Hotel information technology adoption: The case of the Philippines. *ASEAN Journal on Hospitality and Tourism*, 4, pp. 1–15.

Egger, R. (2013). The impact of near field communication on tourism. *Journal of Hospitality and Tourism Technology*, 4(2), pp. 119–133.

Entre (2018). How information technology has affected the tourism and hospitality industry. *Entremt.com*. Retrieved from: https://www.entremt.com/information-technology-affected-tourism-hospitality-industry/ (accessed: the 16th November, 2021).

EU-Vietnam Business Network. (2018). *E-Commerce Industry in Vietnam 2018*. Ho Chi Minh City: EU-Vietnam Business Network.

Falk, M. (2014). *A Panel Data Analysis of Foreign Direct Investment in the Hospitality Industry*. Retrieved from: http://encuentros.alde.es/anteriores/xviieea/trabajos/f/pdf/219.pdf (accessed: the 16th November, 2021).

Floricic, T. and Pavia, N. (2017). Technology innovations as drivers to hotel attractiveness. In O. Dvouletý, M. Lukeš and J. Mísař (Eds.), Proceedings of the 5th International Conference on Innovation Management, Entrepreneurship and Sustainability (IMES 2017). (pp. 231–248). Prag, Češka: Vysoká škola ekonomická v Praze.

Geisler, R. (2018). *Artificial Intelligence in the Travel & Tourism industry Adoption and Impact*. MA Dissertations. Lisboa: Universidade Nova de Lisboa.

Gül, M., Gül, K., Åêîíîìiêà, Đ. and Ekohomika, P. (2018). The effect of fourth industrial revolution on tourism. *Economic Bulletin of the Ukrainian State Chemical-Technological University*, 2(8), pp. 33–37.

Haley, U. C. V. and Haley, G. T. (1997). When the tourists flew in: Strategic implications of foreign direct investment in Vietnam's tourism Industry. *Management Decision*, 35(8), pp. 595–604.

Happ, É. and Ivancsó-Horváth, Z. (2018). Digital tourism is the challenge of future: A new approach to tourism. *Knowledge Horizons: Economics*, 10(2), pp. 9–16.

Härting, R. C., Reichstein, C., Härtle, N. and Stiefnomics, J. (2017). Potentials of digitization in the tourism industry: Empirical results from German experts. In W. Abramowicz (Ed.), Proceedings of the International Conference on Business Information Systems (pp. 165–178). Poznan: 28th-30th, June, 2017.

Hof, A. and Blázquez-Salom, M. (2015). Changing tourism patterns, capital accumulation, and urban water consumption in Mallorca, Spain: A sustainability fix? *Journal of Sustainable Tourism*, 23(5), pp. 770–796.

Hojeghan, S. B. and Esfangareh, A. N. (2011). Digital economy and tourism impacts, influences and challenges. *Procedia Social and Behavioural Sciences*, 19, pp. 308–316.

Ivanov, S. (2019). Ultimate transformation: How will automation technologies disrupt the travel, tourism and hospitality industries? *Zeitschrift für Tourismuswissenschaft*, 11(1), pp. 25–43.

Ivanov, S. and Webster, C. (2017a). Adoption of robots, artificial intelligence and service automation by travel, tourism and hospitality companies: A cost-benefit analysis. In International Scientific Conference "Contemporary Tourism: Traditions and Innovations". Sofia, Bulgaria: Sofia University: 19th–21st October, 2017.

Ivanov, S. and Webster, C. (2017b). Designing robot-friendly hospitality facilities. In Proceedings of the Scientific Conference "Tourism. Innovations. Strategies" (pp. 74–81). Bourgas: 13th–14th October, 2017.

Ivanov, S. and Webster, C. (2019). Robots in tourism: A research agenda for tourism economics. *Tourism Economics*, 26(7), pp. 1065–1085.

Jokonya, O. (2019). Investigating factors influencing ICT adoption among SMEs in the hospitality industry in Western Cape. In Digital Innovation and

Transformation Conference: Digital Skills 2019 Shifting the Digital Skills Discourse for the 4th Industrial Revolution (pp. 83–96). Gauteng: 29th August, 2019.

Kiprutto, N., Kigio, F. W. and Riungu, G. K. (2011) Evidence on the adoption of E-tourism technologies in Nairobi. *Global Journal of Business Research*, 5(3), pp. 55–66.

Kobres, E. (2018). New technologies will revolutionize the hospitality industry. *Forbes.* Retrieved from: https://www.forbes.com/sites/forbestechcouncil/2018/06/28/new-technologies-will-revolutionize-the-hospitality-industry/ (accessed: the 19th November, 2021).

Koetsier, J. (2020). COVID-19 Accelerated E-Commerce Growth '4 to 6 Years'. *Forbes.* Retrieved from: https://www.forbes.com/sites/johnkoetsier/2020/06/12/covid-19-accelerated-e-commerce-growth-4-to-6-years/?sh=28d2845d600f (accessed: the 9th November, 2021).

KPMG in India and FICCI (2018). *Expedition 3.0: Travel and Hospitality Gone Digital: Innovations, Challenges and Way Forward to Win the Indian Consumer.* Retrieved from: https://home.kpmg/in/en/home/insights/2018/03/ficci-expedition-travel-hospitality-technology-innovation-india-digital.html (accessed: the 11th November, 2021).

Law, R., Leung, D., Au, N. and "Andy" Lee, H. (2013). Progress and development of information technology in the hospitality industry: Evidence from Cornell Hospitality Industry: Evidence from Cornell Hospitality Quarterly. *Cornell Hospitality Quarterly*, 54(1), pp. 10–24.

Lu, T. L. (2019). *Opportunities of Artificial Intelligence in Hospitality Industry for Innovative Customer Services Case: Hotels in Ho Chi Minh City, Vietnam.* Bachelor's Thesis. Rovaniemi: Lapland University of Applied Sciences.

Lubbe, B. and Douglas, A. (2019). *An Assessment of the Impact of Mega Trends in the Tourism Sector Technology Trends.* Pretoria: University of Pretoria.

Lyons, J. (2017). How technology is changing the hotel industry and the impact on guest experiences. *Rentokil.* Retrieved from: https://www.rentokil.com/blog/hotel-industry-technology/#.XoB5Y4gzbIU (accessed: the 19th November, 2021).

Mariani, M. (2020). Big data and analytics in tourism and hospitality: A perspective article. *Tourism Review*, 75(1), pp. 299–303.

MarutiTech (2020). *How is the Hotel Industry using AI to provide an awesome User Experience? MarutiTech.* Retrieved from: https://marutitech.com/hotel-industry-ai-awesome-user-experience/ (accessed: the 18th November, 2021).

Menon, N. (2016). ASEAN's digital economy. *Asian Management Insights*, 3(2), pp. 28–33.

Mondejar, R., Chu, C. W. L. and Gogatz, A. (2021). *The Effects of Covid-19 on Digitalization in Southeast Asia: A Crisis within a Crisis.* Retrieved from: https://www.asiaglobalinstitute.hku.hk/researchpaper/effects-covid-19-digitalization-southeast-asia-crisis-within-crisis (accessed: the 19th November, 2021).

Na, S. A., Nee, A. Y. H. and Onn, C. Y. (2017). Medical tourism: The effects of perceived benefits, perceived risks and geographic region. *Pertanika Journal of Social Sciences &. Humanities (JSSH)* 25(Special), pp. 143–152.

Naishtein, R. (2018). Commentary: Beneath the digital friendly façade, Singaporeans still reluctant to accept cashless payments. *Channel News Asia.* Retrieved from: https://www.channelnewsasia.com/news/commentary/singapore-cashless

-society-digital-e-payment-smart-nation-facade-10937012 (accessed: the 29th November, 2021).

Nam, S. (2018). *Assessing the impacts of Foreign Direct Investment (FDI) on Local Skills Development: The Hotel Industry in Siem Reap, Cambodia.* Master of International Tourism Management Thesis. Auckland: Auckland University of Technology.

Nguyen, H. (2019). *Google Reports Highest Mobile Internet Use in ASEAN Region.* Voice of America. Retrieved from: https://www.voanews.com/silicon-valley-technology/google-reports-highest-mobile-internet-use-asean-region (accessed: the 29th November, 2021).

Nonthapot, S. (2017). Causality between capital investment in the tourism sector and tourist arrivals in ASEAN. *Journal of Advanced Research in Law and Economics*, 8(30), pp. 2504–2511.

Nunkoo, R. and Seetanah, B. (2018). Foreign direct investment and tourism development: A theoretical and empirical review. In C. Cooper, S. Volo, W. C. Gartner and N. Scott (Eds.), *Sage Handbook of Tourism Management.* London: Sage, pp. 556–568.

Ong, L. X. I. (2010). Can self service technologies work in the hotel industry in Singapore? A conceptual framework for adopting self service technology. In *Master of Hospitality Administration Project.* Singapore: National University of Singapore.

Online Reporters (2013). Hotel investment boom in SE Asia. *Bangkok Post.* Retrieved from: https://www.bangkokpost.com/thailand/politics/347717/hotel-investment-rise-in-myanmar-cambodia-and-vietnam (accessed: the 19th November, 2021).

Organisation for Economic Co-operation and Development (OECD). (2017). *OECD Southeast Asia Regional Forum: Opportunities and Policy Challenges of Digitalisation in Southeast Asia.* Retrieved from: www.oecd.org/southeast-asia/events/regional-forum/Forum_Note_Digital_Transformation_STI.pdf (accessed: the 25th November, 2021).

Oskam, J. and Boswijk, A. (2016). Airbnb: The future of networked hospitality businesses. *Journal of Tourism Futures*, 2(1), pp. 22–42.

Pathak, M. (2015). Five ways hotels can use technology for a better guest experience. *eHotelier.com.* Retrieved from: https://insights.ehotelier.com/insights/2015/07/20/5-ways-hotels-can-use-technology-for-a-better-guest-experience/ (accessed: the 25th November, 2021).

Perić, J. andRadić, N. M. (2016). Development impact of FDI in hotel: Case study of terme Tuhelj in Croatia. *Alma Tourism: Journal of Tourism, Culture and Territorial Development*, 7(13), pp. 66–78.

Phocuswire (2019). *How Smart Tourism Developments in Thailand Can Help Attract Visitors.* Retrieved from: https://www.phocuswire.com/amadeus-thailand-tourism (accessed: the 12th November, 2021).

Pindţo, R. and Barjaktarović, L. (2018). Digital transformation of tourism. Tourism in function of development of the Republic of Serbia: Tourism in the era of digital transformation. In 3rd International Scientific Conference (pp. 340–355). Vrnjacka Banja: 31st May–2nd June, 2018.

Pratap, A. (2021). Factors affect demand in the tourism sector. *Notesmatic.com.* Retrieved from: https://notesmatic.com/factors-affecting-demand-in-the-tourism-sector/ (accessed: the 12th November, 2021).

Purbasari, R., Muttaqin, Z. and Sari, D. S. (2021). Digital entrepreneurship in pandemic Covid 19 era: The digital entrepreneurial ecosystem framework. *Review of Integrative Business and Economics Research*, 10(Supplementary Issue 1), 114–135.

Radi´c, M. N. (2018). Terrorism as a determinant of attracting FDI in tourism: Panel analysis. *Sustainability*, 10(12), p. 4553.

Record, R. J. L., Larson, B. R., Teh Sharifuddin, S. B. and Chong, Y. K. (2018). *Malaysia's Digital Economy: A New Driver of Development* (English). Working Paper No. 129777. Retrieved from: documents.worldbank.org/curated/en/435571536244480293/MalaysiasDigital-Economy-A-New-Driver-of-Devel opment. (accessed: the 25th November, 2021).

Revfine (n.d.b). *8 Examples of Robots Being Used in the Hospitality Industry*. Retrieved from: https://www.revfine.com/robots-hospitality-industry/ (accessed: the 6th November, 2021).

Revfine (n.d.a). *Hospitality Trends: The Latest Trends in the Hospitality Industry*. Retrieved from: https://www.revfine.com/hospitality-trends/ (accessed: the 6th November, 2021).

Ribeiro, R. and Florentino, T. (2016) Digital transformation in tourism: A high level analysis of the impact that social networks and mass collaboration concept is having at tourism service providers. In 7th International Conference on Cinema and Tourism–ICCT. Porto: 19th–20th October, 2016.

Sabinezs (2018). Effects of industry 4.0 on tourism industry: Development from E-tourism to M-tourism. *Wyred*. Retrieved from: https://wyredproject.eu/2018/04/17/effects-of-industry-4-0-on-tourism-industry-development-from-e-tourism-to-m-tourism/ (accessed: the 12th November, 2021).

Salazar, A. (2018). Hospitality trends: Opportunities and challenges. *Worldwide Hospitality and Tourism Themes*, 10(6), pp. 674–679.

Salleh, N. Z. M., Hamid, A. B. A., Hashim, N. H. and Omain, S. Z. (2010). Issues and challenges in Malaysian hotel operations. In Proceeding of 3rd Asia-Euro Tourism, Hospitality and Gastronomy Conference. Subang: Taylor's College.

Saratchandran, V. (2018). 6 Ways hotels can use technology to improve guest experience. *Fingent.com*. 31 December 2018. Retrieved from: https://www.fingent.com/blog/how-hotels-are-using-technology-for-competitive-advantage (accessed: the 11th November, 2021).

Sathirathai, S. and Wan, M. (2018). ASEAN e-commerce: Hidden tigers meet tech dragons. *The Nation Thailand*. Retrieved from: www.nationthailand.com/business/30338037 (accessed: the 4th November, 2021).

Saulat, A. (2018). Four ways AI is re-imagining the future of travel. *Mindtree.com*. Retrieved from: https://www.mindtree.com/blog/four-ways-ai-re-imagining-future-travel (accessed: the 4th November, 2021).

Sawada, Y. and Sumulong, L. R. (2021). *Macroeconomic Impact of COVID-19 in Developing Asia*. ADBI Working Paper 1251. Retrieved from: https://www.adb.org/publications/macroeconomic-impact-covid-19-developing-asia (accessed: the 5th November, 2021).

Scoop (2020). *Brunei on Track for 2019 Tourism Target, Sees 16% YOY Growth*. Retrieved from: https://thescoop.co/2020/01/16/brunei-on-track-to-meet-2019- tourism-target/ (accessed: the 5th November, 2021).

Siam Commercial Bank Economic Intelligence Centre (SCBEIC) (2017). *Three Megatrends to Change the Face of the Thai Tourism Industry*. Retrieved from:

https://www.scbeic.com/en/detail/file/product/3368/eol77tgr4j/EIC_Insight_tourism_2017_EN.pdf (accessed: the 16th November, 2021).

Singapore Business Review (SBR) (2018). *Hotel Rankings: Smart Hotels Are on the Rise.* Retrieved from: https://sbr.com.sg/hotels-tourism/exclusive/hotel-rankings-'smart-hotels-are-rise (accessed: the 14th November, 2021).

Site Minder (n.d.). *What Do Guests Say About the 10 Most High-tech Hotels in the World?* Retrieved from: https://www.siteminder.com/r/trends-advice/hotel-travel-industry-trends/10-worlds-high-tech-future-hotels/ 1/7 (accessed: the 5th November, 2021).

Snyman, J. A. (2007). *A Foreign Direct Investment Model for Tourism Property Acquisition.* PhD Thesis. Potchefstroom: North-West University.

Soava, G. (2015). Development prospects of the tourism industry in the digital age. *Young Economists' Journal*, 25, pp. 101–116.

Statista.com (2022). *Internet User Penetration in Thailand from 2017 to 2020 with a Forecast Through 2026.* Retrieved from: https://www.statista.com/statistics/975067/internet-penetration-rate-in-thailand/ (accessed: the 5th April, 2022).

Stauvermann, P. J. and Kumar, R. R. (2017). Productivity growth and income in the tourism sector: Role of tourism demand and human capital investment. *Tourism Management*, 61, pp. 426–433.

Stubbs, M. (2018). 5 ways technology is changing consumer expectations in hospitality. *eHotelier.* Retrieved from: https://insights.ehotelier.com/insights/2018/01/23/5-ways-technology-changing-consumer-expectations-hospitality/ (accessed: the 5th November, 2021).

Suriyankietkaew, S. and Nimsai, S. (2021). COVID-19 impacts and sustainability strategies for regional recovery in Southeast Asia: Challenges and opportunities. *Sustainability*, 13, p. 8907.

Tang, T. C. and Wong, K. N. (2011). Foreign direct investment, merchandise and services trade in a transition economy: The case of Cambodia. *International Economic Journal*, 25(2), pp. 251–267.

Technologue.id (2020). *Lawan Corona, Kominfo Apresiasi Platform Digital Karya Anak Bangsa.* Retrieved from: https://kominfo.go.id/content/detail/25534/lawan-corona-kominfo-apresiasi-platform-digital-karya-anak-bangsa/0/sorotan_media, April 02, 2020. (accessed: the 5th November, 2021).

The ASEAN Post Team (2018). E-wallets could boost the region's tourism industry. *The ASEAN Post.* Retrieved from: https://theaseanpost.com/article/e-wallets-could-boost-regions-tourism-industry (accessed: the 9th November, 2021).

Thom, T. (2019). Foreign investors eye hotels in Vietnam: JLL. *The Saigon Times.* Retrieved from: https://english.thesaigontimes.vn/72618/foreign-investors-eye-hotels-in-vietnam-jll.html (accessed: the 24th November, 2021).

Truong, V. D. and Le, A. (2016). The Evolution of Tourism Policy in Vietnam, 1960-2015. In C. M. Hall and S. J. Page (Eds.), *The Routledge Handbook of Tourism in Asia.* London: Routledge, pp. 191–204.

United Nations Conference on Trade and Development (UNCTAD) (2007). *FDI in Tourism: The Development Dimension.* Retrieved from: unctad.org/en/Docs/iteiia20075_en.pdf (accessed: the 4th November, 2021).

Vandenberg, P. (2020). *How SMEs Can Bounce Back from the COVID-19 Crisis. Development Asia Policy Brief.* Manila: Asian Development Bank.

Velikova, E. (2019). Innovation and digitalization in tourism: Restriction or development for business in Bulgaria. *Medical Journal of Trakya University/Trakya Universitesi Tip Fakultesi Dergisi*, 17(1), pp. 252–258.

Viet An Law Firm (2017). Foreign investment in Cambodia up 800% in 10 years. *B2BCambodia.com.* Retrieved from: https://www.b2b-cambodia.com/news/foreign-investment-in-cambodia-up-800-in-10-years/10/07/2017 (accessed: the 16th November, 2021).

Vietnam Economic News (VEN) (2016). *Hospitality Attracts Foreign Investors.* Retrieved from: ven.vn/apicenter@/print_article&i=1480 ½ (accessed: the 7th November, 2021).

Vietnamplus (2019). *Tourism Administration, VNPT Develop Smart Tourism.* Retrieved from: https://en.vietnamplus.vn/tourism-administration-vnpt-develop-smart-tourism/152262.vnp (accessed: the 7th November, 2021).

Widiatedja, I. G. N. P. (2019). FDI on tourism and the environmental damage: What Indonesia's trade and investment agreements can do? *Yuridika*, 34(2), pp. 303–324.

Widiatedja, I. G. N. P. and Wairocana, I. G. N. (2018). Fragmented and unclear laws and regulations of foreign direct investment in Indonesian tourism. *Padjajaran Journal of Law*, 5, pp. 408–409.

World Bank (2020). *Individuals Using the Internet (% of Population) – Vietnam.* Washington, DC: The World Bank. Retrieved from: https://data.worldbank.org/indicator/IT.NET.USER.ZS?locations=VN (accessed: the 7th November, 2021).

World Economic Forum (WEF) (2017). *Digital Transformation Initiative: Aviation, Travel and Tourism Industry.* Cologny: WEF.

Xu, S., Stienmetz, J. and Ashton, M. (2020). How will service robots redefine leadership in hotel management? A Delphi approach. *International Journal of Contemporary Hospitality Management*, 32(6), pp. 2217–2237.

Yeo, S. (2019). How tech can help Southeast Asia keep up with surging tourist arrivals. *TechInAsia.com.* Retrieved from: https://www.techinasia.com/tech-southeast-asia-surging-tourist-arrivals 1/5 (accessed: the 16th November, 2021).

Part 2
Examples

3 Blockchain Technology in Aviation, Tourism, and Hospitality

Theoretical and Practical Understanding

Anis Najiha Ahmad and NurHusna Samsudin

Introduction

Tourism is one of the major economic sectors. The sector is a main driver of economic growth and job creation. In 2019, the World Travel and Tourism Council (WTTC) reported that tourism supported about 300 million jobs; this is equivalent to 1 in 10 jobs. Comparatively, the total contribution of the tourism industry in 2019 accounted for 10.4% of the total global gross domestic product (GDP) (World Travel and Tourism Council, 2019). The tourism industry consists of different interactive and connected sectors such as travel, transportation (e.g., aviation), and hospitality (e.g., accommodation, restaurants, catering services, and entertainment). The terms "travel," "tourism," and "hospitality" are often used interchangeably. In fact, services provided by the travel and hospitality industry are partly geared towards catering to tourism activities and partly for other non-tourism purposes. The components of the tourism industry are shown in Figure 3.1. Each component in the industry is supported by large and different businesses and public agencies that include the aviation and hospitality sectors.

The growth of the tourism industry is directly reflected in the growth of the aviation and hospitality sectors. Air transport, for instance, is the main mode of transport for international tourists. In 2017, it is estimated that 54% of international tourists used this mode to reach their destinations. In terms of job opportunities, it is estimated that over 36 million jobs in the aviation sector, out of a total of 62.7 million jobs, are connected to the tourism industry. This translates to over $892 billion out of the total $2.7 trillion in worldwide economic activity (Economic, 2017). The hospitality sector is responsible for providing essential services (i.e., accommodation and food) for tourists. In 2017, the hospitality industry accounted for 313 million jobs worldwide, which translates to 9.9% of total employment and 20% of all global net jobs created in the past decade (World Travel and Tourism Council, 2018).

The tourism industry, like other industries, is continuously exploring innovative solutions to strengthen its competitive advantage, reduce costs, improve customer satisfaction, and improve overall organizational efficiency and performance. As one of the latest and most innovative technologies, there is a growing interest in the ways blockchain can be leveraged in the tourism industry. So far, blockchain

DOI: 10.4324/9781003315179-6

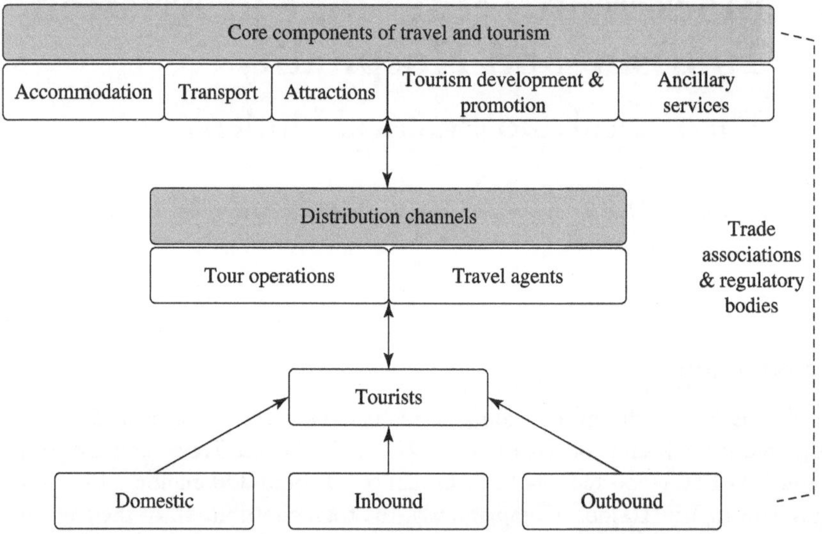

Figure 3.1 Components of the tourism industry (source: Dale, 2007).

has been strategically used within the global airline industry and also the broader travel industry, including in the accommodation, travel agent, and food service sectors. However, the massive adoption of blockchain is still slow.

The chapter is structured as follows. The next section presents the fundamental underlying technology of blockchain and its characteristics. The following section identifies and discusses the potential applications of blockchain technology in the tourism industry, specifically in the aviation and hospitality sectors. The challenges of potentially adopting this innovative technology are discussed afterward. In a nutshell, this chapter contributes to enhancing the understanding of the current state of the art, benefits, applications, and adoption challenges of applying blockchain technology in the tourism industry.

Concept of Blockchain

The basic concept of blockchain has its core in the pioneer cryptocurrency platform, Bitcoin. The technology enables a digital currency to create a reliable transaction network without depending on a principal authority and to operate as a transaction ledger that keeps identical copies of all transactions across all network members' computers (Yaga et al., 2018). Due to the fact that the ledger is dispersed over the network, each stakeholder will have access to the past and most recent entries, which aids in the blockchain's security (Peters et al., 2015; Yaga et al., 2018). Members of the network are anonymous, non-trusted individuals called nodes. Each node may participate in the general agreement process, but a limited number of nodes are tasked with confirming the data or transaction that will be integrated into the block (Yaga et al., 2018). The information could be a person's private identity, work history, healthcare records, etc. A consensus in the network is required to verify a new piece of information or transaction before it is added to the chain. Once enclosed in a block, all the information is permanently recorded (Peters et al., 2015).

How Does Blockchain Technology Function?

When a new block is created, it is added to the blockchain. A blockchain is a collection of linked blocks. Four conditions must be met before a block may be added to the blockchain (Reiff, 2020). The first aspect is that a transaction must occur. As an illustration, consider a hypothetical purchase made on an internet platform. After promptly clicking a checkout button and making a purchase, a block will gather possibly thousands of transactions. The purchase will be included in the block along with the information about other users' transactions. Following that, the transaction must be validated. In contrast to public data records, which need someone to oversee the verification of new data inputs, the blockchain delegates that responsibility to a network of computers. When someone makes an online purchase, that network of computers confirms the particular purchase and ensures that the transaction went as planned (Reiff, 2020). One of the most critical components of blockchain technology is the way transactions are confirmed and validated. To understand this validation concept, one should know an important term called "private key cryptography." To demonstrate the technology behind private cryptographic keys, consider two people wanting to execute an online transaction. Each of these people possesses two keys: one is secret, while the other is public. This element of cryptography enables individuals to build a safe digital identification reference point by merging their public and private keys. A critical component of blockchain technology is this safe identity (Zheng et al., 2017). A public and private key pair together to form a digital signature, which is an effective tool for validating and regulating ownership.

As in the prior example, if two individuals want to conduct an online transaction and each has a private and public key, blockchain allows the first individual (person A) to use their private key to associate transaction data with the second individual's public key (person B). This data is merged to create a block, which contains a digital signature, a timestamp, and other transaction-related data, but not the identities of the parties involved in the transaction. This block is then sent to all nodes, or other network components, that serve as validators for the transaction through the blockchain network (Zheng et al., 2017).

The third aspect is that, after a transaction has been accurately verified, it gets permission to be stored in a block. The information, like the amount of the transaction and the digital signature of both parties, the buyer and seller, can all now be stored in a block. There, that transaction will join a thousand other purchases. Lastly, the block must be given a name, or in the digital world, it is referred to as a hash. A hash is a unique, identifying code that is given to the block after the transaction has been verified (Yaga et al., 2018). Additionally, the block will be hashed against the most recent block added to the chain. Once the block has been given a hash, it can be added to the blockchain. Once it is put on the blockchain, it becomes publicly accessible, allowing anybody to read the transaction details. Figure 3.2 displays the flow of how the blockchain works (Buterin, 2015).

Science of Blockchain Systems

The three types of blockchain systems now in use are public blockchains, private blockchains, and consortium blockchains. All records in a public blockchain are available to the public, and anybody can participate in the consensus process. In the case of private blockchains, only nodes belonging to a single organization would be permitted to participate in the consensus process. In contrast, a consortium blockchain's consensus process would involve just a small number of pre-selected nodes (Hyperledger Foundation, n.d.).

Due to the fact that a private blockchain is entirely controlled by one company, it is considered a centralized network. The consortium blockchain developed by many organizations is somewhat decentralized, as only a subset of nodes is used to determine consensus (GitHub Inc., n.d.). The following section compares the three types of blockchains.

Consensus determination: In a public blockchain, any node can participate in the consensus process. And with a consortium blockchain, just a subset of nodes is responsible for verifying the block. In the case of a private chain, it is entirely controlled by a single entity that has the ability to select the ultimate consensus.

Read permission: In a public blockchain, all transactions are accessible to the public. However, in a private blockchain or a consortium blockchain, visibility varies.

Immutability: Due to the distributed storage of data, it is practically impossible to tamper with transactions on a public blockchain. By comparison, transactions on a private blockchain or a consortium blockchain can be readily manipulated due to the small number of participants.

Efficiency: The propagation of transactions and blocks takes a long time due to the high number of nodes on the public blockchain network. As a result, transaction data is constrained, and latency is significantly increased. Consortiums and private blockchains may be more efficient with fewer validators.

Centralized: The primary distinction between public, consortium, and private blockchains is that public blockchains are decentralized, consortium blockchains are somewhat centralized, and private blockchains are completely centralized since they are owned by a single party.

Consensus process: Anyone on the planet may participate in the public blockchain's consensus process. Both consortium and private blockchains, in contrast to public blockchains, are permissioned.

(a) Each confirmed block references the previous one, forming a chain of blocks.

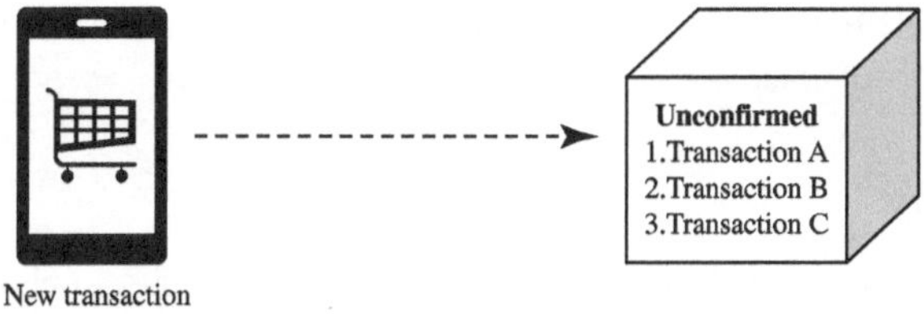

(b) Before being added to the block, information (transaction) is pending (unconfirmed)

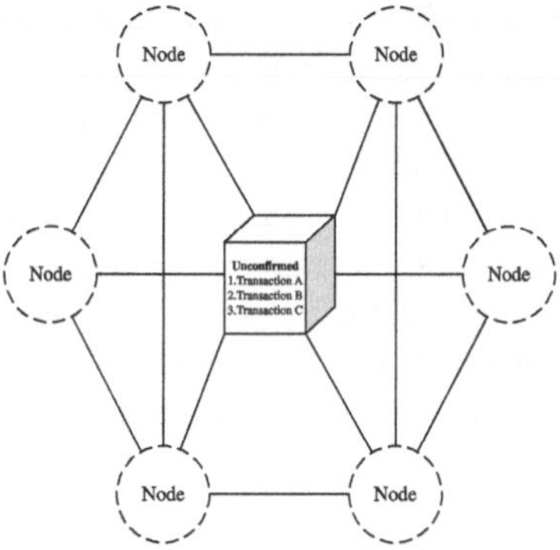

(c) Details of unconfirmed transactions are distributed across the entire network.

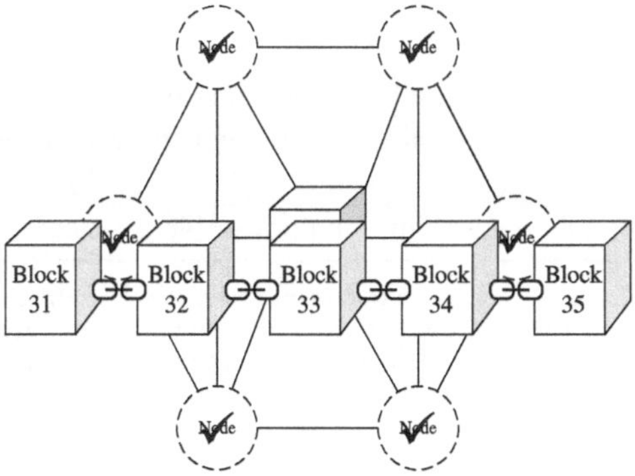

(d) This block is sent to all other nodes in the network. All nodes check and verify that the block is correct, and by consensus, add it to their copy of the chain.

Figure 3.2 Flow of how the blockchain works (source: Buterin, 2015).

Because public blockchains are available to the whole globe, they can attract a large number of users and communities to engage actively. Numerous public blockchains are emerging over time. As for consortium blockchains, they have the potential to be utilized in a wide variety of commercial applications. For instance, Hyperledger is actively building blockchain frameworks for corporate consortiums. Additionally, Ethereum has developed tools for constructing consortium blockchains (GitHub Inc., n.d.). Table 3.1 simplifies the comparison of public, private, and consortium chains.

Blockchain Privacy

While anybody may view the data on the blockchain, consumers can choose to connect their computers to the network as nodes. As a result, each computer gets a copy of the blockchain that is updated in real time whenever a new block is released, similar to how a Facebook newsfeed changes in real time whenever a new status is made (Hyperledger Foundation, n.d.).

Each computer connected to the blockchain network keeps a copy of the blockchain, resulting in thousands of identical copies. While each copy of the blockchain is identical, the data is more difficult to manipulate when distributed over a network of computers. With blockchain, there is no one, unalterable record of events (Dang, 2015). Even while blockchain transactions are not completely anonymous, the person initiating the transaction remains anonymous. The only information that users provide is their digital signature or username. A hacker would have to compromise every copy of the network's blockchain instead. This is what it means when a blockchain is referred to as a "distributed" ledger. As a consequence, the issue of blockchain system security and trust was brought to the forefront.

Blockchain Security System

Security and trust challenges are addressed in a number of ways by blockchain technology. To begin with, new blocks are always stored in a linear and chronological

Table 3.1 Comparison of public, consortium, and private blockchains

Property	*Public chain*	*Consortium blockchain*	*Private blockchain*
Consensus determination	All miners	Selected set of nodes	One organization
Read permission	Public	Could be public or restricted	Could be public or restricted
Immutability	Nearly impossible to tamper with	Could be tampered with	Could be tampered with
Efficiency	Low	High	High
Centralized	No	Partial	Yes
Consensus process	Permissionless	Permissioned	Permissioned

(Source: Jagjit, 2021.)

fashion. That is, they are always attached to the blockchain's end. After a block is added to the end of the blockchain, it is very difficult to modify its contents. This is due to each block containing its own hash in addition to the previous block's hash (Reiff, 2020). Hash codes are created by converting digital information to a string of numbers and letters using a math function. If that data is altered in any way, the hash code is altered as well (Reiff, 2020). For instance, suppose a hacker attempts to edit an online store transaction in such a way that the buyer is forced to pay twice for the purchase. Long before they adjust the dollar value of the transaction, the block's hash will change. The subsequent block in the chain will preserve the prior hash, and the hacker must alter that block to cover their trail (Reiff, 2020). Nevertheless, doing so modifies the block's hash.

A hacker would thus need to edit every following block on the blockchain in order to alter a single block. Calculating all of those hashes would take an absurd amount of processing power. In other words, once a block is put on the blockchain, it becomes exceedingly difficult, if not impossible, to change or delete it (Reiff, 2020). To address the trust problem, blockchain networks have included verification mechanisms for machines desiring to join the chain and contribute blocks. The tests, termed "consensus models," mandate that users "prove" their authenticity before joining a blockchain network. The network will need a computer to show that it has completed work by successfully solving a tough computational arithmetic problem (Moser, 2013). If a computer resolves one of these problems, it is eligible to add a block to the blockchain. However, the process of adding blocks to the blockchain, which the cryptocurrency industry refers to as "mining," is not straightforward. While proof of work does not prevent hacker attacks, it does make them less effective. To carry out a blockchain attack, a hacker would need to control more than 50% of the network's processing power in order to overwhelm all other members (Zheng et al., 2017).

Blockchain Applications in the Tourism Industry

Blockchain technology is already well entrenched and expanding in cryptocurrency and financial services. It has now started to find its way into other fields such as healthcare, supply chains, risk management, internet of things (IoT), and public and social services (Zheng et al., 2018). Tourism products are highly service-oriented and intangible, relying on reputation management and trust. This one characteristic of the tourism industry aligns well with the capabilities of blockchain technology, given trust is the basis of blockchain technology (Shin, 2019).

Since the technology is still in its infancy, comprehensive application of the technology in tourism is rather limited (Amadeus, 2017). Nevertheless, the applications are expanding and perhaps are better adopted in the tourism industry compared to other industries. For example, the Société Internationale de Télécommunications Aéronautiques (SITA) (2018), an IT provider for the air transport industry, reported that, in 2018, 59% of airlines surveyed had invested in pilot studies or research programmes related to blockchain. Nowadays, blockchain technology enables the management and administration of the tourism industry throughout the chain via tools such as smart contracts, decentralized applications, and cryptocurrencies (Ozdemir et al., 2019).

Blockchain technology applications in the aviation and hospitality sectors are discussed in the following sub-sections. Multiple applications of blockchain have been identified. Many companies in the hospitality and aviation industries have started to embrace and apply blockchain technology to improve the efficiency of their processes, reduce costs, and improve customers' experiences. Blockchain technology enables a secure means of storing data and conducting transactions, ensuring that data cannot be manipulated. Blockchain technology may also provide a safe and secure method of processing and utilizing the numerous permanent digital documents required by both the aviation and hospitality sectors. This concept may also be applied to a variety of other facets of the sectors, including preserving security and identity, ticketing, loyalty programmes, aviation maintenance, baggage management, and online review platforms. It is also useful for the authorities to regulate the tourism sector.

Payments

The main application of blockchain technology is to enable easy, fast, secure, and traceable payments not only in tourism but in all other industries or fields as well. Payments with cryptocurrency enable more secure and traceable transactions because all transactions carried out on the blockchain are recorded on the chain and cannot be modified. The decentralized system of the blockchain means there will be no intermediaries to intervene or delay the payment. Using cryptocurrency exchange, international travellers are free from the volatility of foreign exchange operations (Accenture, 2020).

Security and Identity

An attack on the information technology (IT) infrastructure of airline companies that consists of customer and operational data (onboard systems, and core networks) can severely affect airline operations and result in financial losses. In some cases, the breaches may result in security implications whereby the safety of staff and passengers could be jeopardized. Given the sensitive nature of airline data, the protection of data security and privacy is highly important. The use of blockchain technology would provide for a lower risk manner of managing and sharing information across various entities (Accenture, 2020).

Ticketing

In most airports and airlines throughout the world, electronic ticketing systems have mostly replaced conventional paper tickets. An airline's current electronic ticketing system uses a database entry that is connected with the airline's passenger service system. The database is linked to all other partners, including airports, ground transportation, and travel companies, to enable real-time data exchange (Crosby, 2007). Blockchain could help improve the current reservation system for airlines or their ticketing system by tokenizing this asset and further dematerializing it. Blockchain could offer a better, secure, transparent, and stable ticketing system. Using the smart contracts tool that is connected with the asset, blockchain may be used to add the necessary business logic, terms, and conditions around how the ticket is sold and used (Accenture, 2020). This will affect the way airline tickets are distributed and purchased. The blockchain technology enables airlines to strengthen their alliances by enabling their partners to sell tickets in real time, regardless of location. In addition, according to Eventmb Studio Team (2019), a blockchain ticket platform is also more efficient in handling high-demand tickets.

Loyalty Programme

A loyalty programme is a marketing strategy to retain customers by rewarding them for their repeat purchase behaviour. It is an incentive programme to encourage returning customers, or the loyal customer. Airline companies, for instance, offer frequent-flyer programmes (FFP) also known as loyalty programmes. Besides airlines, other travel service providers such as hotels and restaurants also run their own loyalty programmes. Traditionally, customers must accumulate and wait until they have enough points before using them. There are also restrictions on where the points can be spent (Accenture, 2020). However, customers nowadays are not really interested in accumulating points but are keener on faster, personalized, and engaging rewards. This makes the traditional mode of loyalty programmes less attractive to customers.

The use of blockchain could potentially help the tourism industry revolutionize its loyalty programmes. Tokenized loyalty points on the blockchain allow customers to check and redeem the points instantly. Tokenization is the "process of transformation of rights to an asset into a digital token on a blockchain" (Tokend, 2018). As blockchain allows the loyalty points to be used as "currency" among more providers, customers could choose to redeem rewards with the business partner on the platform that accepts the token (Accenture, 2020; Tokend, 2018). In addition, cryptocurrency can also be used during future transactions as it has no expiry date (Joshi, 2019). Blockchain-based loyalty programmes are also tougher to hack. As the points are tokenized, unique identifiers are created that are extremely difficult to counterfeit. More than 51% of its nodes must be hacked to access information recorded on a blockchain. This provides multiple levels of security that were not possible to offer previously (Fromhart and Therattil, 2016). Services like Trippki (a hotel booking and rewards platform) use a blockchain-based loyalty programme for their customers. Trippki facilitates direct interaction between the customer and hotels, allowing the hotels to offer their visitors the best service and deals (Joshi, 2019).

Aviation Maintenance

Blockchain technology could also potentially transform the way maintenance logs or records are processed. Currently, there are two ways the aviation industry maintains its logs or records: using electronic records (EAMR) or manual/paper-based records. Although the EAMR records are better than the manual, paper-based records, multiple digital databases could cause inefficiencies, time-consuming retrieval processes, and a lack of transparency. In addition, electronic records are also easily exposed to data manipulation as they are stored in centralized databases. In the case of an audit or accident, the investigation could take time as each stakeholder maintains their own database. As access to aircraft maintenance records is limited to specific maintenance agencies, manufacturers, or airline companies are not granted easy access to the data. This is a slow process of information exchange (Takyar, 2020).

Blockchain technology could overcome the limitations of the current digital databases and manual paper-based records by creating an immutable digital record of each maintenance activity in public and private databases. The records include the maintenance logs (e.g., frequency, person in charge), manufacturing details, component details, and other sensitive information about the aircraft (Accenture, 2020; Takyar, 2020). In addition, the technology could potentially be used to verify the authenticity of the parts supplied in plane maintenance. Using smart contracts to manage how the data flows within the blocks through the chain, all stakeholders could access the maintenance records (Takyar, 2020; Fernandez, 2019). This, in turn, improves the security of maintenance activities and data exchange transparency (Accenture, 2020).

Baggage Management

Baggage mishandling and loss are common concerns for airline companies and customers. Baggage is moved multiple times during international flights and may be misplaced in transit. Besides affecting the reputation of the airlines, baggage loss also has a subsequent cost implication for the companies, as the airlines are required to compensate passengers for damaged, delayed, or lost baggage. Baggage loss or mishandling can be significantly reduced using blockchain and other technologies (barcoded tags and radio-frequency identification (RFID)-enabled tags). A combination of blockchain technology, AI, and sensor technologies can be used to identify and monitor luggage. For instance, sensors attached to bags help track the exact location of luggage even if the handlers misplace it.

Coordination and Management

One of the most promising features of blockchain is the ability to reduce third party mediation, thus promoting immediate provider-to-consumer interaction. The intermediary plays a defined role in the structured stages of the tourism chain. Intermediaries are the dealers who link the chain of distribution between the companies (hotels, airlines, restaurants) and their customers (tourists) (see Figure 3.3). Intermediaries include multiple suppliers of travel products such as travel agents, tour operators, wholesalers, bed banks, booking centres, destination management companies (DMCs), and online travel agencies (OTAs) (Lewis et al., 1998). In the past, before the internet, customers had limited ability to directly access tourism products. Customers had to reserve and go through a "chain" of intermediaries, who were paid a commission for their services (see Figure 3.4) (Kracht and Wang, 2010). Airlines, for instance, were dependent on travel agents as an intermediary to sell their seats. In the year 2000, approximately 80% of all sales tickets were done through travel agents (Lubbe, 2000). The advance of the internet has not reduced the number of intermediaries in the distribution channel as expected, but rather transformed the chain and structure into a more complex array of intermediaries (Kracht and Wang, 2010).

As such, decentralized management systems via blockchain technology could potentially further reduce the tourism industry's dependency on intermediaries. Using blockchain, information flows securely while being accessible, reducing management times, and providing universal access to information for all stakeholders. Winding Tree is an example of a real-life blockchain application. This company is developing a decentralized trip reservation platform with the aim of eliminating the external middlemen and significantly reducing the price by up to 20% (Vidal, 2020).

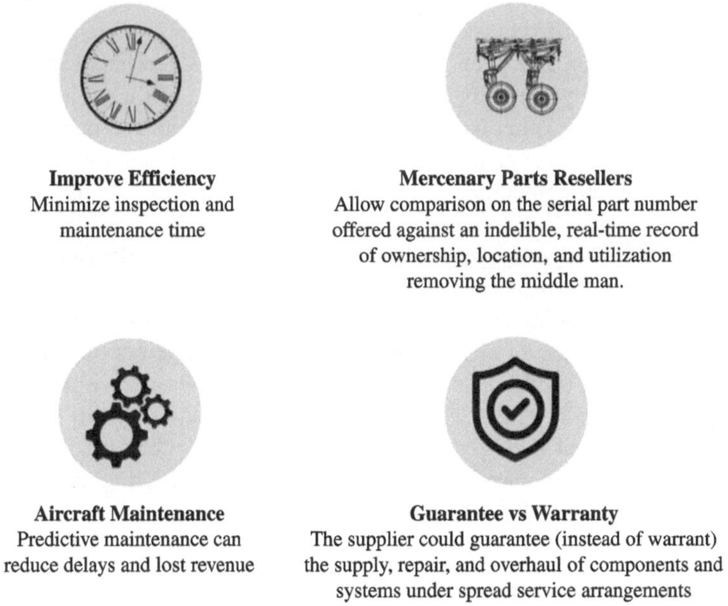

Figure 3.3 The benefits of a blockchain platform for airplane maintenance (source: PwC, 2019).

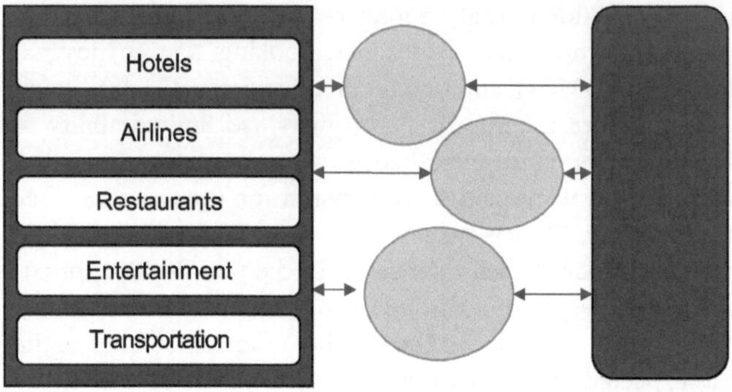

Figure 3.4 Intermediaries linked tourism service providers with tourists (source: Dragan et al., 2015).

Online Review Platforms

Consumer-generated content (CGC), such as online review platforms like TripAdvisor and Expedia, is increasingly used as a guide to inform travel planning. These platforms are popular as they are considered to be more credible and natural than biased promotional campaigns (Bickart and Schindler, 2001). Because of the significance of online reviews, businesses in the tourist sector are under pressure to obtain and retain positive word of mouth (Xu et al., 2019). However, the reviews on these platforms are prone to "fraud reviews" as irresponsible parties can pose as real customers and write "consumer" reviews. Fake reviews can be positive or negative. Companies may post fake positive reviews of themselves to yield financial gains. On the other hand, negatively reviewed companies could be affected in terms of reputation damage and monetary loss. The unsecure online platform allows unethical companies to write fake negative reviews about their competitors.

Because all the information that appears on the network is public, dependable, and secure, blockchain could protect the credibility of review platforms. Information stored on the blockchain is highly secure. This enables greater transparency and consumer trust. One example of a real-world application of online reviews using the blockchain platform is the CoolCousin travel company. It uses a decentralized peer-to-peer community to connect travellers directly to someone they can trust. The platform enables travellers to get personalized services and information from locals. The method removes the creation of false rating profiles by rating websites using complicated artificial intelligence (AI) engines.

Regulatory Authorities

The applications of blockchain technology have a wide range of potential benefits, not only for businesses but also regulators. Standard implementation of blockchain could potentially improve the ability of the authorities to ensure that operators and service providers comply with regulations and standards. In addition, if a blockchain-based electronic personnel licencing platform becomes a standard, regulatory authorities could easily verify the authenticity of pilots and aircraft through reliable and transparent record keeping in the aviation context.

The Challenges for Potential Adoption of Blockchain

Despite the promising potential of blockchain technology, there are a myriad of challenges and risks that could diminish this potential. A new technological tool can be a double-edged sword. Potential barriers and challenges to the adoption of blockchain are outlined below. These include possible privacy breaches, standardization issues, lack of experts, legal uncertainty, technological concerns, and customer financial incentives.

Privacy

Although promoted as secure and privacy-preserving data management and storage, privacy and confidentiality are still one of the concerns of blockchain adoption (Casino et al., 2019). Some have highlighted that sensitive data could be accessed and compromised in the case of improper management, as the information in blockchain is accessible and stored in a public ledger. Previous research has shown that blockchain transaction history can be linked to a personal identity (Smith and Khovratovich, 2016).

Standardization

Experts perceive a lack of standardization as a substantial barrier to blockchain technology's mass adoption (Aristidou and Marcou, 2019; Zainuddin, 2019). Fundamentally, blockchain technology relies on creating links and interoperable networks within the tourism industry. Although blockchain serves as the base technology, different organizations could create their own networks and algorithms, which makes communication between different systems or stakeholders tricky. The lack of such uniformity across blockchain protocols has become a barrier to entry for new developers, professionals, and investors. Inconsistency in basic processes like security makes mass adoption unlikely. Standardization introduces interoperability. It helps develop more efficient consensus mechanisms to improve operational efficiencies, which also reduces costs (Zainuddin, 2019).

Shortage of Experts

Another challenge to the further development of blockchain is the shortage of talent in technical skills. These experts required include developers, security experts, and business consultants (Akmeemana, 2017; Casino et al., 2019). Besides technical expertise, legal expertise is also crucial in order to design, build, implement, and, lastly, operate blockchain systems within the new business models. Legal experts must have a precise understanding of the programming language associated with many blockchain platforms and smart contracts. This requirement has proven to be a challenge for many. It is reported that law firms are having difficulty keeping up with the cryptocurrency and blockchain industry's demand for lawyers (Alexandre, 2019). The shortage of expertise is also linked to other hindrances in the implementation due to the cost of expertise and resources (Casino et al., 2019).

Legal Uncertainty

For years, experts have voiced their concerns about the legal and regulatory barriers, or specifically the lack of regulatory clarity for the application of blockchain technology (Zainuddin, 2019). Given its novelty, blockchain technology applications may bring uncertainties. These uncertainties represent a risk to the public as the full array of blockchain's eventual applications is unknown at this point. For example, the validity and legal enforceability of smart contracts are still unclear. It can be quite difficult for the government to forecast the potential dangers that they pose to the public. As such, experts are cautious about how government agencies may interact with blockchain technology (Kehoe and Hallahan, 2018). Therefore, all stakeholders need to be abreast of regulatory developments and continuously engage with policy makers (PwC, 2019).

Technology Concerns

The core concept of blockchain is still in its early stages. Advancements in functionality and usability are still in progress. There are also concerns regarding the immutability of data in the system (Kehoe and Hallahan, 2018).

Ecosystem Participants' Reluctance to Adopt Blockchain

Just like with other novel technologies, more time is needed to leverage blockchain to its full potential. Bringing the whole industry ecosystem into the blockchain realm requires complex coordination (Smith and Khovratovich, 2016). As blockchain is a network-based technology, a majority of market participants need to join the bandwagon to fully realize its value and potential. This may take years (Kehoe and Hallahan, 2018).

Customer Financial Incentives

As proven in other new technology adoptions (e.g., electric cars), a lack of financial incentives is one of the significant barriers for customers to adopt the technology.

In the blockchain case, while the financial incentives or cost reduction benefits for companies in replacing the intermediary are evident, the benefits in terms of cost saving from a customer's point of view are less obvious. The replacement of the intermediary would not automatically guarantee that the price for the customer would be reduced (Calvaresi et al., 2019). This factor may hinder the mass adoption of blockchain among customers.

Conclusion

Blockchain technology, one of the finest cutting-edge technologies of the 21st century, has the potential to dramatically transform the tourism industry. Given the characteristics and possibilities offered by blockchain, it could maximize the security, accessibility, and efficiency of the tourism value chain. There is an impending need for tourism stakeholders, industry, and academics to closely monitor the development of blockchains and explore their applications to leverage the technology's potential. It is important to note that, while there's consensus about its potential, it is still difficult to know where it can get to and its real impact on the industries, as the technology is still in the early stages of its life.

References

Accenture (2020). *The Potential of Blockchain for Airlines.* Retrieved from: https://www.accenture.com/us-en/insight-potential-blockchain-technology-for-airlines (accessed: 7th October, 2021).

Akmeemana, C. (2017). *Blockchain Takes Off-how Distributed Ledger Technology Will Transform Airlines.* Blockchain Research Institute. Retrieved from: https://www.brightline.org/resources/blockchain-takes-off-how-distributed-ledger-technology-will-transform-airlines/ (accessed: 19th October, 2021).

Alexandre, A. (2019). Demand for legal experts with blockchain tech knowledge is high. *Cointelegraph.* Retrieved from: https://cointelegraph.com/news/demand-for-legal-experts-with-blockchain-tech-knowledge-is-high-report (accessed: 2nd October, 2021).

Amadeus. (2017). *Blockchain: Harnessing Its Potential in Travel.* Retrieved from: https://amadeus.com/en/insights/research-report/blockchain-harnessing-its-potential-in-travel (accessed: 12th December, 2021).

Aristidou, C. and Marcou, E. (2019). Blockchain standards and government applications. *Journal of ICT Standardization*, 7(3), pp. 287–312.

Bickart, B. and Schindler, R. M. (2001). Internet forums as influential sources of consumer information. *Journal of Interactive Marketing*, 15(3), pp. 31–40.

Buterin, V. (2015). On public and private blockchains. *Ethereum Foundation Blog.* Retrieved from: https://blog.ethereum.org/2015/08/07/on-public-and-private-blockchains/ (accessed: 2nd October, 2021).

Calvaresi, D., Leis, M., Dubovitskaya, A., Schegg, R. and Schumacher, M. (2019). Trust in tourism via blockchain technology: results from a systematic review. In J. Pesonen and J. Neidhardt (Eds.), *Information and Communication Technologies in Tourism 2019* (pp. 304–317). Cham: Springer International Publishing.

Casino, F., Dasaklis, T. K. and Patsakis, C. (2019). A systematic literature review of blockchain-based applications: current status, classification and open issues. *Telematics and Informatics*, 36, pp. 55–81.

Crosby, T. (2007). How airline E-tickets work. *howstuffworks*. Retrieved from: https://adventure.howstuffworks.com/destinations/travel-guide/tips/how-airline-e-tickets-work.htm (accessed: 18th October, 2021).

Dale, G. (2007). *Travel and Tourism*. London: Pearson Education.

Dang, Q. A. (2015). The Bologna process goes east? From "Third Countries" to prioritizing inter-regional cooperation between the ASEAN and EU. In A. Curaj, L. Matei, R. Pricopie, J. Salmi and P. Scott (Eds.), *The European Higher Education Area: Between Critical Refections and Future Policies* (pp. 763–783). Cham: Springer International Publishing.

Dragan, D., Kramberger, T. and Topolšek, D. (2015). Supply chain integration and firm performance in the tourism sector. In Pre-Conference Proceedings of the 12th International Conference on Logistics & Sustainable Transport 2015. Celje, Slovenia: University of Maribor, Faculty of Logistics.

Economic (2017). Aviation and tourism: the economics. *Aviation: Benefits Beyond Borders*. Retrieved from: //aviationbenefits.org/blog/2017/01/aviation-and-tourism-the-economics/ (accessed: 2nd October, 2021).

Eventmb Studio Team (2019). *Blockchain for Ticketing: A Complete Guide*. Retrieved from: https://www.eventmanagerblog.com/blockchain-ticketing (accessed: 2nd October, 2021).

Fernandez, A. (2019). How blockchain could enhance aircraft maintenance? *Datascience.aero*. Retrieved from: https://datascience.aero/blockchain-enhance-aircraft-maintenance/ (accessed: 24th October, 2021).

Fromhart, S. and Therattil, L. (2016). *Making Blockchain Real for Customer Loyalty Rewards Programs*. Deloitte. Retrieved from: https://www2.deloitte.com/content/dam/Deloitte/us/Documents/financial-services/us-fsi-making-blockchain-real-for-loyalty-rewards-programs.pdf (accessed: 2nd October, 2021), pp. 1–17.

GitHub Inc. (n.d.). *Ethereum*. Retrieved from: https://github.com/ethereum/wiki/wiki/Consortium-Chain-Development (accessed: 2nd October, 2021).

Hyperledger Foundation (n.d.). *Learn*. Retrieved from: https://www.hyperledger.org/ (accessed: 9th October, 2021).

Jagjit, S. (2021, November 3). *Private, Public and Consortium Blockchains: The Differences Explained, Cointelegraph the Future of Money*. Retrieved from: https://cointelegraph.com/explained/private-public-and-consortium-blockchains-the-differences-explained.

Joshi, N. (2019). How blockchain can transform tourism. *Allerin*. Retrieved from: https://www.allerin.com/blog/how-blockchain-can-transform-tourism (accessed: 2nd October, 2021).

Kehoe, L. and Hallahan, J. (2018). *Blockchain: A Game Changer in Aircraft Leasing?*. Deloitte. Retrieved from: https://www2.deloitte.com/content/dam/Deloitte/ie/Documents/Tax/Airfinance%20Journal%201.pdf (accessed: 2nd October, 2021).

Kracht, J. and Wang, Y. (2010). Examining the tourism distribution channel: evolution and transformation. *International Journal of Contemporary Hospitality Management*, 22(5), pp. 736–757.

Lewis, I., Semeijn, J. (Jake). and Talalayevsky, A. (1998). The impact of information technology on travel agents. *Transportation Journal*, 37(4), pp. 20–25.

Lubbe, B. (2000). The changing role of the travel intermediary. *South African Journal of Economic and Management Sciences*, 3(2), pp. 273–289.

Moser, M. (2013). Anonymity of bitcoin transactions: an analysis of mixing services. In Proceedings of Munster Bitcoin Conference. Munster, Germany, pp. 17–18.

Ozdemir, A. I., Ar, I. M. and Erol, I. (2019). Assessment of blockchain applications in travel and tourism industry. *Quality and Quantity*, 54, pp. 1549–1563.

Peters, G. W., Panayi, E. and Chapelle, A. (2015). Trends in crypto-currencies and blockchain technologies: a monetary theory and regulation perspective. *SSRN Electronic Journal.* Retrieved from: https://doi.org/10.2139/ssrn.2646618 (accessed: 15th October, 2021).

PwC (2019). *How Blockchain in Aerospace Can Improve Efficiency.* Retrieved from: https://www.pwc.com/gx/en/industries/aerospace-defence/publications/blockchain-in-aerospace.html (accessed: 27th October, 2021).

Reiff, N. (2020). *Bitcoin vs. Ethereum: What's the Difference?. Investopedia.* Retrieved from: https://www.investopedia.com/articles/investing/031416/bitcoin-vs-ethereum-driven-different-purposes.asp (accessed: 26th October, 2021).

Shin, D. D. H. (2019). Blockchain: the emerging technology of digital trust. *Telematics and Informatics*, 45, 101278.

Smith, S. M. and Khovratovich, D. (2016). Identity system essentials. *Evemyrn.* Retrieved from: https://www.evernym.com/wp-content/uploads/2017/02/Identity-System-Essentials.pdf (accessed: 29th October, 2021).

Société Internationale de Télécommunications Aéronautiques (SITA) (2018). *Air Transport IT Insights 2018.* Retrieved from: https://www.sita.aero/resources/type/surveys-reports/air-transport-it-insights-2018 (accessed: 11th October, 2021).

Takyar, A. (2020). *Blockchain Aviation: Blockchain Use Case.* Retrieved from: https://www.leewayhertz.com/blockchain-aviation-better-transparency-trust/ (accessed: 2nd October, 2021).

Tokend (2018). *Tokenization for Loyalty Programs.* TokenD. Retrieved from: https://tokend.io/loyalty/ (accessed: 12th October, 2021).

Vidal, B. (2020). What we can expect from blockchain technology in tourism. *We Are Marketing.* Retrieved from: //www.wearemarketing.com/blog/what-we-can-expect-from-blockchain-in-the-tourism-industry.html (accessed: 27th October, 2021).

World Travel and Tourism Council (WTTC) (2018). *Travel and Tourism Economic Impact 2018.* London: WTTC.

World Travel and Tourism Council (WTTC) (2019). *Insights and Reports.* Retrieved from: https://wttc.org/en-gb/Research/Insights (accessed: 22nd October, 2021).

Xu, Y., Zhang, Z., Law, R. and Zhang, Z. (2019). Effects of online reviews and managerial responses from a review manipulation perspective. *Current Issues in Tourism*, 27(13), pp. 2207–2222.

Yaga, D., Mell, P., Roby, N. and Scarfone, K. (2018). *Blockchain Technology Overview.* Retrieved from: https://doi.org/10.6028/NIST.IR.8202 (accessed: 18th October, 2021).

Zainuddin, A. (2019). Challenges in blockchain technology today. *eToroX.* Retrieved from: https://www.etorox.com/news/opinions/challenges-in-blockchain-technology-today/ (accessed: 22nd October, 2021).

Zheng, Z., Xie, S., Dai, H., Chen, X. and Wang, H. (2017). An overview of blockchain technology: architecture, consensus, and future trends. In 2017 IEEE International Congress on Big Data (BigData Congress) (pp. 557–564). Honolulu, HI, USA: IEEE. The 25th–30th June, 2017.

Zheng, Z., Xie, S., Dai, H.-N., Chen, X. and Wang, H. (2018). Blockchain challenges and opportunities: a survey. *International Journal of Web and Grid Services*, 14(4), pp. 352–375.

4 Artificial Intelligence, Biometric Technology, and Mobile Technology Application in Aviation and Tourism

How It Changes the Landscape of Tourism and Air Travel

Mohd Khir Harun, Ahmad Zahir Mokhtar, Reshminder Kaur Satvindar Singh, and Nor Aida Abdul Rahman

Introduction

The fourth industrial revolution has led to rapid changes in the application of technology in many sectors including aviation, air travel, and tourism. Technology usage in the aviation sector has improved efficiency as well as travellers' experience. According to Pew Research Centre (2019), technology is rapidly changing and about 5 billion people own a mobile device and 2.65 billion people use social media and mobile technology. In fact, it is reported that artificial intelligence (AI) technology will produce US$2.9 trillion of business value worldwide. In the aviation context, for example, the use of AI technology elevates work efficiency and increases speed, workload, and safety as well as improving navigation, the data ecosystem, and data safety. In the context of travel and tourism, the role of technology has significantly improved travellers' experience especially in booking, travel planning, safety, directions, and ensuring the traveller can stay connected with family and loved ones (Rahman et al., 2020).

After 30 years of the computer and the Internet, there is widespread availability of big data and mobile technology, as well as biometric technology (Mahroof, 2019). These new technologies have played a big role in speeding up the process of globalization. In fact, in the logistics and transport context for example, technology has made the process of transportation and warehousing systems more efficient in terms of moving containers, tracking and tracing, information sharing, ordering and controlling, and much more.

As highlighted by Rahman, Rahman, and Hassan (2020), technology also has improved the landscape of the travel and tourism sector. Recent developments in mobile technology, as well as biometric technology applications, for instance have improved the facilitation process at the airport, saving time and easing the process for travellers. Mobile technology, which mostly refers to cellular

DOI: 10.4324/9781003315179-7

communication, has led to better flexibility for travellers during their tourism activities. With mobile technology applications, travellers can get information at any time and anywhere, they are able to make payments wirelessly, they have improved connectivity with their network and families during travel, they may use the global positioning system (GPS) to reach their desired location, and they can use instant messaging, gaming, and much more. All of these technology benefits show the better quality and flexibility of travellers' activities during travel. In addition, with Internet technology becoming more and more mature, people no longer have to worry about information leakage and information delay.

The popularity of the digitalization era and technology revolution has led to significant progress in the industry especially in improving daily operational activities, decision-making processes, business performance, safety and security, business communication, resource management, and time and money savings, as well as knowledge transfer. According to the latest report by Mlitz (2021) on the global technology forecast, the global market for industry spending on information technology (IT) is estimated to be approximately US$5.3 trillion. Asia, for example, is expected be one of the largest contributors to the global information industry at 31%.

With regard to recent developments in technology, there are many tools and devices which connect to security devices for instance mobile technology, artificial intelligence, and biometric technology. All of these technologies help society to keep interacting, feel safer, and make the world a better place to live. On the other hand, for business players, new technology helps them to be effective and achieve efficiency in utilizing their resources. For instance, AI technology and machine learning have been used in the aviation industry especially in the airline sector to increase the digital interaction between passengers and the airline organization (Clarke and Smith, 2012). Additionally, biometric technology which is largely used for verification purposes in the aviation sector also helps airport operators to facilitate their passengers more effectively (Miller and Fauve, 2012).

Having this background, this chapter aims to clearly discuss the role of the three new technologies, namely artificial intelligence, biometric, and mobile technology, from an air travel and tourism perspective. The discussion will cover the history and recent developments of the technology and how these three technologies have changed the landscape of the air travel and tourism sector. The arrangement of the chapter starts with AI technology and biometric technology, followed by mobile technology. Past studies on these three technologies are also explored. Following that, a number of future studies are also presented as a conclusion to this chapter.

Artificial Intelligence Technology Developments: Multi Sector in Air Travel and Tourism

Artificial intelligence was first proposed in 1956, and it has been used to replace the abilities of human intelligence. The area of AI technology during that time included mathematical logic, computer science, and also computer neurobiology. After 60 years since its emergence, current AI technology has been expanded to machine learning, neural networks, and many others. According to Li, Yin, Qiu, and Bai (2021), the success of AI is connected to heuristic searches, facial recognition systems, processing of natural language, mobile robotics, and many more. After 60 years of its existence, this innovative invention of AI technology has spread to many sectors for multiple purposes such as to build robots and help in decision-making processes. AI is able to analyse, provide, and produce information and knowledge in a user-friendly form which leads to value creation in business operations (Makridakis, 2017; Yoon and Lee, 2019).

A growing corpus of literature has been examining AI technology in multiple sectors such as in the healthcare and pharmacy industry (Smith and Miller, 2021), political science (Saari, 2020), banking (Ali et al., 2019), manufacturing (Ooi et al., 2018), and farming (Kshetri, 2020), as well as in tourism (Pillai and Sivathanu, 2020), the water industry (Soh et al., 2018), and the aviation sector (Talley, 2020). From the aviation and air travel perspective, there has been an increased interest in investigating the role of AI technology from different domains such as prediction, neural networks, machine learning, and algorithms. It is recognized that AI technology could have some vital results in term of decision-making processes, business activity, and also operations. In the field of tourism for instance, the use of AI technology has become ubiquitous during the COVID-19 pandemic to reduce social interaction among other reasons. AI technology such as chatbot has replaced human jobs and decreased the level of human-to-human interaction.

At present, AI technology has changed the landscape of travel and tourism activities. Figure 4.1 highlights the main features of AI technology in the air travel and tourism sector.

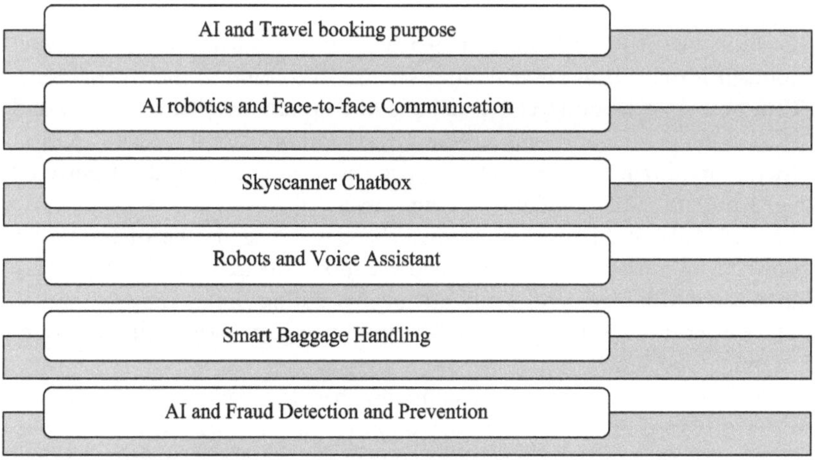

Figure 4.1 The main features of AI technology in the air travel and tourism sector (source: the authors).

AI technology has made the air travel and tourism industry more reliable and attractive. One of the main key features of AI in the travel and tourism sector is for booking purposes. At present, travellers experience communication with machines which replace the role of travel agents. For instance, the chatbot that is deployed in social media such as in WhatsApp, Facebook, Instagram, Skype, and many more is now taking the place of the travel agency especially in accommodation and flight booking activities. If previously, travellers were limited to travel agency packages, at present, travellers have the opportunity of a personalized booking experience.

Secondly, AI technology could also allow the robot to replace the human in serving food and also conveying information. For instance, in medical tourism activity in certain countries, it can be seen that robots are serving food to patients in the hospital. Additionally, at the airport for example, travellers are guided by robots at the air terminal. London's Heathrow Airport is one of the best examples of the use of robots to handle passengers at a busy air terminal. AI has changed the traveller experience landscape; travellers or passengers at the airport no longer need to queue to get information at a counter. This not only reduces the traffic at the airport counter, but is also beneficial to travellers in saving their time.

Thirdly, it can be seen that AI technology such as the Skyscanner chatbot helps travellers to plan their journey and make reservations with hotels. For example, giant travel reservation websites such as Expedia, Skyscanner, or Booking.com use chatbots in communicating with their customers. The bot will communicate with the customer depending on the customer request and helps the customer to get the best booking sites with the best deals and processes.

The next example is robots and voice assistants. AI technology has successfully provided contactless mechanisms to travellers especially during the COVID-19 pandemic. Voice assistants were used for security reasons, while airport for instance are equipped with robotic and AI technology for example in handling lost luggage cases smartly.

The fifth role of AI technology that is changing the landscape of air travel and tourism is smart baggage handling systems. According to the International Air Transport Association (IATA), millions of bags are handled every year and the number of bags is increasing in line with the number of travel and tourism activities. For instance, Eindhoven Airport has successfully used AI-powered luggage handling systems without baggage labels.

The sixth role of AI technology that changes the landscape of travel and tourism activity is fraud detection and prevention. AI-driven technologies are design smartly to identify and verify documents during online transactions. As a result, AI benefits and protects travellers from being involved with fraud cases during online activities.

It has been proven that AI technology has gradually made significant improvements in tourism activities and improved all manual processes before. As highlighted by Samala, Katkam, Bellamkonda, and Rodriguez (2020), AI technology has the ability to change the landscape of air travel and tourism by offering various information and outperforming human services. The next section will discuss biometric technology, which is also connected to the AI technology domain.

Biometric Technology Developments: Multi Sector in Air Travel and Tourism

The word biometric refers to body identification based on biological characteristics. Biometrics, which originally comes from "Greek" word, is a technology used in identifying a person based on certain factors such as fingerprint, face, vascular, DNA, voice, and iris (Thakkar, 2020). In the air travel sector, biometric technology is widely used at airports for identity screening. The method of biometric technology application is quite straightforward. According to Boussadia (2009), biometric technology is suitable for authentication purposes by identifying certain physical and behavioural attributes. Behavioural attributes here refer to lip movement, signature, keystroke, and walk step (gait). These are connected to a process of matching and verifying one's physical characteristics with one's identity (Gillis, n.d.). As highlighted by the IATA (2021), there is one programme which is called "One ID" that allows the traveller or passenger to enjoy their tourism journey with a document-free process. This One ID programme is being developed with an improved identification process using biometric technology. This is one of the latest developments of biometric technology in the context of air travel and tourism. With this, passengers will be able to identify themselves at each airport touchpoint through a simple biometric recognition. The objective is to achieve truly interoperable system coordination between airports, airlines, and governments (Patel, 2018).

As highlighted by Thakkar (2020), there are six types of biometric characteristics used in matching one's identity and personality. Figure 4.2 shows the six

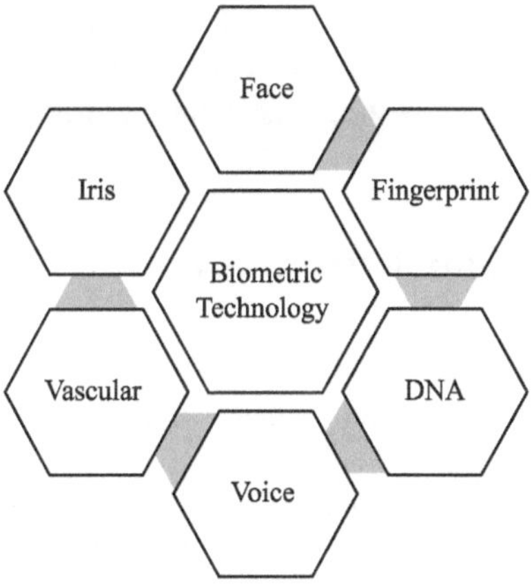

Figure 4.2 Six types of biometric technology (source: the authors).

types of biometric technology applications that have changed the landscape of the air travel and tourism industry.

According to Callahan (2019) in a *Forbes* report, the popularity of biometric technology applications is increasing. Face biometric technology, also known as facial biometric technology, is widely used and accepted in the air travel and tourism industry. It was reported that the global facial recognition market is forecasted to be worth US$6.49 billion by 2021, up from US$2.35 billion in 2016, and China is investing heavily in facial recognition technology. One of the benefits of biometric technology in the air travel and tourism sector is cost and time saving. At the airport, for instance, travellers have to undergo security checks and a number of documents checks at customs and immigration counter, as well as for check in and at the boarding gate. Fingerprint identification technology is also a popular biometric technology application in the context of aviation, travel, tourism, and hospitality. For example, thumbprints were used in collecting digital images for many transactions as it is the most common sensor technology used in many other sector including banking.

Iris or iris scanning is the process of using visible and near-infrared light to take a high-contrast photograph of a person's iris. It is a form of biometric technology in the same category as face recognition and fingerprinting. On the other hand, fingerprints were used early in biometric technology. Vascular biometric technology is also one of the popular methods for identifying and verifying an individual's identity using the vein pattern of one's palm. Voice is also another popular method for identifying the biological characteristics of an individual. In the travel and hospitality industry, it is widely used, for instance, voice biometric applications have been used in the hotel industry to improve customer experience and avoid stolen credentials when logging into the portal online. DNA is an increasingly useful biometric, and is encountered most often in forensics and healthcare. For forensics, current DNA identification technologies measure short tandem repeat sequences (STRs) in the nuclear or mitochondrial DNA. This explanation shows that biometric technology is not only used for the sake of traveller or tourist comfort, but rather the main reason is more connected to the safety and security issues of travellers or passengers, as well as keeping the airport activities safe.

Mobile Technology Developments: Multi Sector in Air Travel and Tourism

The history of mobile technology begins as early as the 1970s when the first mobile phone was introduced. Mobile technology could be defined as portable technology that follows wherever the user goes. In other words, it is a technology that goes where the user goes. At present, the impact of mobile technology applications in the context of aviation, air travel, tourism, and hospitality is substantial. Mobile technology consists of two-way communication devices and the networks that connect them. As technology advances, various mobile device technologies are becoming more promising and popular among travellers. In the

COMMUNICATION	DATA EXCHANGE	SENSORS
Camera	Bluetooth	Barometer
Speaker	Infrared	Magnetometer
Microphone	NFC	Gyroscope
Radio	Personal Hotspot	Touchscreen
Cellular Phone	USB	Gravity
SMS / MMS	IrDA	Proximity
	(FM) RDS	Ambient Temperature & Pressure
	Asissted GPS	Gravity
		Compass
		Relative Humidity

Figure 4.3 Three hardware components of mobile technology development (source: the authors).

future, mobile technology will fundamentally change travel, tourism, and hotel services. As stressed by Kim and Kim (2017), mobile technology has changed the landscape of the travel, tourism, and hospitality environment. Mobile technology allows smart tourism activity to occur such as providing travellers with a rich onsite experience. It creates value for the travellers' journey by providing online data to travellers pertaining to physical infrastructure, destination, food and accommodation availability, social connection, family connection, and much more. It enhances the value added activity of travellers during their journey from one location to another (Li et al., 2017). Additionally, all of these benefits from mobile technology are gathered from mobile device hardware components which consist of sensors, communication, and data exchange. Figure 4.3 illustrates the three components that are the main driving forces for the rapid development of mobile technology in the travel and tourism sector.

Conclusion

This chapter focuses on the development of three key technologies in the context of aviation, tourism, and hospitality. The chapter focus on current developments in AI, mobile, and biometric technology in the travel and tourism sector while focusing on past and current developments that have changed the landscape of the industry. Although there are many other types of technology applications in tourism, such as big data, cyber security, analytics, and others, to provide detailed and focused discussion, these three technologies were selected. Future research is called for to further explore how the technology has improved industry performance empirically. In fact, a review paper that connects the study of technology and related theoretical underpinnings is also required. A review of the underpinning theory used in technology-related research in the area of tourism and aviation will assist future research in having a deep understanding of the theoretical background of the related research. To conclude, more research opportunities are available in the area of technology usage and its impact in the aviation, travel, tourism, and hospitality context. Grounded research methodology is also key to exploring and having a deep understanding of the issue of technology in the travel and tourism context.

References

Ali, H., Abdullah, R. and Zaini, M. K. (2019). Fintech and its potential impact on Islamic banking and finance industry: A case study of Brunei Darussalam and Malaysia. *International Journal of Islamic Economics and Finance*, 2(1), pp. 73–108.

Boussadia, K. (2009). The evolution of airport screening technology. *Biometric Technology Today*, 17(2), pp. 7–8.

Callahan, J. (2019). *How the Travel Industry is Driving Biometric Security Innovation.* Forbes. Retrieved from: https://www.forbes.com/sites/forbestechcouncil/2019/01/09/how-the-travel-industry-is-driving-biometric-security-innovation/?sh=2dea3eb06572 (accessed: 30th August, 2021).

Clarke, M. and Smith, B. (2012). Impact of operations research on the evolution of the airline industry. *Journal of Aircraft*, 41(1), pp. 62–72.

Gillis, A. S., Loshin, P. and Cobb, M. (n.d.). *Biometrics.* Tech Target Search Security. Retrieved from: https://searchsecurity.techtarget.com/definition/biometrics (accessed: 10th August, 2021).

International Air Transport Association (IATA) (2021). *One ID.* Retrieved from: https://www.iata.org/en/programs/passenger/one-id/ (accessed: 19th December, 2021).

Kim, D. and Kim, S. (2017). The role of mobile technology in tourism: Patents, articles, news, and mobile tour app reviews *Sustainability*, 9(11), p. 2082. doi:10.3390/su9112082

Kshetri, N. (2020). COVID-19 meets big tech. *Computer*, 53(8), pp. 10–13.

Li, M., Yin, D., Qiu, H. and Bai, B. (2021). A systematic review of AI technology-based service encounters: Implications for hospitality and tourism operations. *International Journal of Hospitality Management*, 95, 102930.

Li, Y., Hu, C., Huang, C. and Duan, L. (2017). The concept of smart tourism in the context of tourism information services. *Tourism Management*, 58(2), pp. 93–300.

Mahroof, K. (2019). A human-centric perspective exploring the readiness towards smart warehousing: The case of a large retail distribution warehouse. *International Journal of Information Management*, 45, 176–190.

Makridakis, S. (2017). The forthcoming Artificial Intelligence (AI) revolution: Its impact on society and firms. *Futures*, 90, pp. 46–60.

Miller, D. and Fauve, B. (2012). Mobile e-commerce to drive voice-based authentication. *Biometric Technology Today*, 2012(2), pp. 5–8.

Mlitz, K. (2021). *Global Information Technology Industry Forecast 2019–2022 by Region.* Statista. Retrieved from: https://www.statista.com/statistics/507365/worldwide-information-technology-industry-by-region/ (accessed: 28th November, 2021).

Ooi, K.-B., Lee, V.-H., Tan, G. W.-H., Hew, T.-S. and Hew, J.-J. (2018). Cloud computing in manufacturing: The next industrial revolution in Malaysia? *Expert Systems with Applications*, 93, pp. 376–394.

Patel, V. (2018). *Airport Passenger Processing Technology: A Biometric Airport Journey.* Master's Theses. Daytona Beach: Embry–Riddle Aeronautical University.

Pew Research Centre (2019). *Smartphone Ownership Is Growing Rapidly Around the World, but Not Always Equally.* Retrieved from: https://www.pewresearch.org/global/2019/02/05/smartphone-ownership-is-growing-rapidly-around-the-world-but-not-always-equally/ (accessed: 9th January 2022).

Pillai, R. and Sivathanu, B. (2020). Adoption of AI-based chatbots for hospitality and tourism. *International Journal of Contemporary Hospitality Management*, 32(10), pp. 3199–3226.

Rahman, N. A. A., Rahman, M.S-U-. and Hassan, A. (2020). Tourism and air transport sustainability in Bangladesh: The role of technology. In A. Hassan (Ed.), *Tourism Marketing in Bangladesh: An Introduction*. Singapore: Routledge, pp. 42–50.

Saari, M. (2020). IR 4.0. In parliament: Conceptualising the application of artificial intelligence and machine learning in the parliament of Malaysia's parliamentary questions. *International Journal of Law, Government and Communication*, 5(20), pp. 124–137.

Samala, N., Katkam, B. S., Bellamkonda, R. S. and Rodriguez, R. V. (2020). Impact of AI and robotics in the tourism sector: A critical insight. *Journal of Tourism Futures*, 8(1), pp. 73–87. doi:10.1108/JTF-07-2019-0065

Smith, M. and Miller, S. (2021). The ethical application of biometric facial recognition technology. *AI & SOCIETY*, 37(1), pp. 167–175. doi:10.1007/s00146-021-01199-9

Soh, Y. W., Koo, C. H., Huang, Y. F. and Fung, K. F. (2018). Application of artificial intelligence models for the prediction of standardized precipitation evapotranspiration index (SPEI) at Langat River Basin, Malaysia. *Computers and Electronics in Agriculture*, 144, pp. 164–173.

Talley, S. M. (2020). Public acceptance in the AI technology in self flying aircraft. *Journal of Aviation/ Aerospace Education & Research*, 29(1), pp. 48–64.

Thakkar, D. (2020). An overview of biometric iris recognition technology and its application areas. *Bayometric.* Retrieved from: https://www.bayometric.com/biometric-iris-recognition-application/ (accessed: 10th August, 2021).

Yoon, S. N. and Lee, D. H. (2019). Artificial intelligence and robots in healthcare: What are the success factors for technology-based service encounters? *International Journal of Healthcare Management*, 12(3), pp. 218–225.

5 Big Data Analytics in the Southeast Asian Aviation, Tourism, and Hospitality Industries

Current and Future Developments

Nor Hisham Ghafar

Introduction

Aviation, tourism, and hospitality are different industries but closely and directly inter-related to each other. Aviation is the catalyst for people's movement. The movement of passengers domestically and internationally triggers the needs for accommodation and food, which become the anchor of the hospitality services. When many people travel to-and-fro and stay in a variety of places, these accelerate the demand in tourism as well as the hospitality services. Nevertheless, the increasing popularity of tourism destinations in Southeast Asian regions has increased the demand tremendously for air travel, tourism, and hospitality services.

As a region so rich in diverse and unique multiracial cultures and natural eco-environments, Southeast Asia, which consists of Malaysia, Indonesia, Thailand, Singapore, the Philippines, Vietnam, Cambodia, Brunei, Myanmar, Laos, and Timor-Leste, has become a central attraction location for people all over the world. As depicted in Table 5.1, the number of arrivals to all Southeast Asian countries is increasing with Thailand and Malaysia remaining the most popular attraction points. The increasing number of air travellers every day creates more opportunities as well as challenges in managing those industries strategically.

In recent years, the rapid development of technology has disrupted the traditional trends in the aviation, tourism, and hospitality services. The roles of information and data have now become strategically central inputs for aviation, tourism, and hospitality service providers to offer unique, personalized, and inspirational travel experiences to customers (Yallop and Seraphin, 2020). In other words, in the era of globalization bridged by the digital transformation, big data analytics has become the major enabler of the evolving smart concepts in aviation, tourism, and hospitality (Ardito et al., 2019).

Categories and Sources of Big Data Generation in Aviation, Tourism, and Hospitality

Big data offers two types of information, namely structured and unstructured data. Subjective data gathered from individual posts, photos, and complaints made

DOI: 10.4324/9781003315179-8

Table 5.1 International tourist arrivals in Southeast Asia by country

Country	*Tourists arrivals (in millions)*		
	2017	*2018*	*2019*
Malaysia	25.95	25.83	26.10
Indonesia	12.95	13.40	15.46
Thailand	35.60	38.18	39.80
Singapore	13.90	14.67	15.12
Philippines	6.62	7.17	8.26
Vietnam	12.92	15.50	18.01
Cambodia	5.60	6.20	6.61
Brunei	0.26	0.28	0.32
Myanmar	3.44	3.55	4.36
Laos	3.26	3.77	4.38
Timor-Leste	0.07	0.08	0.08

(Source: United Nations World Tourism Organization, 2020.)

via tourism websites, social media, and blog forums are primarily unstructured data. These data require mining, cleaning, analyzing, and interpreting processes in order to present them as more meaningful information to users. Structured data, on the other hand, are data that can be directly quantified and classified into categories for analysis purposes, such as personal details, flight details, and hotel booking details.

As depicted in Table 5.2, big data generation within the aviation, tourism and hospitality industries can be divided into three sources. The first source of big data generation is through human-generated data. Human-generated data are the data which are directly and personally furnished by a person to the other parties who he/she deals with. Examples of structured human-generated data are personal details given by an airline passenger through form filling or the payment information given during online booking through a hotel's website. Examples of unstructured human-generated data are photos or videos of attractive tourism places and activities uploaded and shared by a traveller on his social media, or complaints made via emails on bad service received from a travel agency.

The second category of big data generation source is device-generated data. Device-generated data refers to any information auto-generated and shared by any internet-connected device or machine after a person uses them. Examples of structured device-generated data are the GPS location auto-detected by a smart phone, automated time zone detected, or time taken by a traveller to travel from home to a destination as auto-generated by Google Maps. Examples of unstructured device-generated data are the passenger's biological conditions detected when using facilities/machines in an airport, or traffic condition auto-alerts by a passenger's smart phone.

Table 5.2 Categories of big data and sources of big data generation

Categories of big data	*Sources of big data generation*		
	Human-generated data	*Device-generated data*	*Operation-generated data*
Structured data	• Personal data given during online flight ticket booking. • Information on payment made to hotel such as name, amount paid, credit card number.	• GPS locations auto-detected by smart phone. • Automated time zone. • Time taken for travellers' journey from home to airport.	• Flight routes as planned by flight operation. • Passengers manifest generated by airline system. • Guest information during actual check-in and -out at hotel's front desk.
Unstructured data	• Photo or video posts on social media by travellers of attractive tourism destinations and experiences. • Comments made on blogs on the quality of hotel services. • Complaint emails written to tourist agency.	• Users' biological conditions when using facilities/ machines in airport. • Traffic condition auto-alert. • Video search preference/ interest in tourism destinations on YouTube.	• Weather conditions shared by air traffic controller to pilot. • Maintenance reports made by pilots to ground crew.

(Source: developed by the authors.)

The final category of big data generation source is operation-generated data. Operation-generated data refers to the information that must be generated and/ or furnished to users due to the operation requirements of air travelling. Examples of structured operation-generated data are the flight routes as planned by the flight operators of the airlines, passengers' manifest, or guests' personal details given to the check-in counter of a hotel. Examples of unstructured operation-generated data are the weather conditions information shared by an air traffic controller with pilots, or the maintenance reports made by a pilot to the ground maintenance crew. Nevertheless, data containing any combination of both structured and unstructured data are categorized as semi-structured data.

Examples of Applications of Big Data Analytics in Major Southeast Asian Countries

Big Data Analytics in Malaysia's Airports Operations

Airports are a sector in aviation which gain benefits from big data analytics. Malaysia Airport Holdings Berhad (MAHB), which is the major airport operator in the country, operating airports including the Kuala Lumpur International Airport (KLIA) and its second terminal KLIA2, capitalizes data analytics technology to enhance passengers' personalized experience through a Unified Digital Interface and a fully fledged analytics platform. The app, known as MYAirports, operates based on real-time digital interaction with passengers. Its enables passengers to plan their journey to the airport ahead of time, and its real-time intelligent navigation feature assists passengers to travel from home, through the checkpoints, to the boarding gate in the shortest time possible. This technology is integrated with other airport systems such as a passenger flow management system, automated queue management system, and customer feedback management system.

MAHB is also developing big data management technology which integrates predictive analytics, service management, and automation management, supported machine learning, and an artificial intelligence system. All data generated by each passenger who uses these digital platforms in the airport will be channelled and stored in a cloud and later analyzed and utilized using artificial intelligence and data analytics techniques to ensure the smooth digital experience of other passengers using the same airport.

Big Data Analytics in Thailand's Tourism and Hospitality Industry

The tourism industry has long been the backbone of the economy for Thailand. Several initiatives have been taken by the Thailand government to boost the attractiveness of local tourism including capitalizing on big data analytics technology. One of the recent efforts is the Tourism Smart Data Management project, a collaboration between the Tourism Authority of Thailand (TAT) and a multinational ride-hailing company, Grab Taxi (Thailand), to promote local tourism (Tourist Authority of Thailand, 2019). Each time tourists use the Grab application for ride-hailing services, all data on tourists' traffic and behaviour are collected anonymously from more than hundreds of thousands of daily rides. Among the collected data are places visited, tourists' spending, tourism activities in each place visited, and travel statistics (Sritama, 2018).

These big data from Grab are then shared with the TAT through the Tourism Smart Data Management system to analyse and generate relevant insights on the behaviour and lifestyle of both international and local tourists, their favourite destinations, and the length of stay, as well as their travelling patterns whilst in Thailand. With this information, TAT works with the most popular search engine company, Google, and the most popular social media platform in Thailand, Facebook, to promote attractive tourism products and attractive travelling packages for potential travellers. TAT also aligns its tourism promotional campaigns with the services offered by local online travel agents, who will then propose the travel routing and deal with the local community's accommodation providers through Airbnb. Such smart collaboration with well-known multinational companies like Grab, Google, Facebook, and Airbnb, via big data analytics technology platforms, has helped Thailand not only to retain repeat travellers but also attract new visitors from all over the world (Vajirakachorn and Chongwatpol, 2017).

Big Data Analytics in Singapore's Hospitality Industry

Singapore is among the early adopters of Industry 4.0 technology including big data analytics technology especially in the hospitality industry. Since the hospitality services generate a huge amount of data from millions of travellers checking in into various hotels every day, the use of big data analytics in the hospitality industry creates opportunities for hotels to understand the preference patterns of customers and subsequently offer personalized packages (Wee et al., 2019). Hilton Hotel Singapore for example has capitalized on big data analytics to gain a 360-degree understanding of its customers' behaviours and preferences. With these useful customer insights, Hilton is able to deliver more value to them, which in turn results in repeat customers and happier guests (Brar, 2019).

At the government level, in order to create better tourist experiences, Singapore Tourism Board (STB) has initiated the Singapore Tourism Analytics Network (STAN) to collect and analyse daily data from tourists who visit Singapore such as spending patterns and lengths of stay in hotels (GovInsider, 2018). Through this information, STB is able to cluster tourist travelling patterns based on nationality and their respective favourite hotels as well as their spending behaviour. For example, STAN's big data analysis reveals that one out of ten tourists who checked in to one hotel would be more likely to switch hotels on the third day if they were visiting Singapore for more than five days. The predictive analysis of STAN also reveals that a majority of Japanese and South Korean tourists who switch hotels on the third day tend to choose a more expensive hotel. Such information is very useful for hoteliers to design more attractive and personalized promotional bundles to the same cluster of tourists to encourage their longer stay in Singapore which eventually contributes to income generation for the country.

Big Data Analytics in Indonesia's Digital Tourism Marketplace

Indonesia is blessed with many cultural and natural resources which attract many tourists from all over the world. For this reason, in 2017, the Indonesian government declared tourism as the fourth among the top five priority sectors for the country. Geographically, Indonesia is the largest archipelagic country consisting of more than 13,000 islands with two-thirds of its area covered by water. These geographical characteristics offer a variety of tourism infrastructure, tourism services, and tourism human resource quality in different tourism destinations. However, connectivity and wide geographical dispersion have been major problems for the government in integrating tourism activities and products. Due to this reason, the Ministry of Tourism Indonesia has launched Indonesia Travel Exchange (ITX) as the digital platform to integrate the different tourist destinations and tourism stakeholders nation-wide (Pranita, 2018).

ITX acts as a digital marketplace by collecting daily data from potential tourists who *look* at, *book*, and *pay* for airline tickets, hotel reservations, restaurants, transport rental services, and event selections before they travel to Indonesia. These data are then used by the tourism stakeholders including 70,000 homestay villages across the archipelago (Yuniarni, 2017) to offer more attractive tailored service and to enhance destination promotion efforts between the tourist agencies and the providers of tourism products.

Future Trends in Big Data Analytics in Aviation, Tourism, and Hospitality

Studies show that big data technology will continuously evolve in the future. With the technology on big data advancing, the trend on relying on data analytics will also be intensified. The aviation industry will be heavily driven by big data when all the equipment, machines, and applications used by the aviation infrastructure and aircraft will have the capability to capture, collect, and store all the data of the humans who interact with or use them. Some airlines, for instance, have been investing heavily in developing their big data analytics capability to support a variety of decisions involving multiple business processes (Hausladen and Schosser, 2020). In addition, the aircraft will also become smarter when many sensors are installed. These can sense any damage while still flying and then channel those data in real-time to the ground maintenance crew even before the same aircraft lands in the airport. This will mean some conventional human maintenance will be replaced with predictive maintenance driven by the massive data generated through those sensors.

In addition, airports are moving towards smart airports where all systems owned by airlines, airports, and cargo will be linked, interacting, and exchanging information with each other. All this information becomes input to improve the passengers' experience physically and digitally when they travel through that airport. The amount of big data generated, collated, and shared by airlines, airports, air traffic control, and passengers will have to be handled by advanced artificial intelligence and autonomous machines with high predictive capability.

The tourism industry is also transforming, aligned with the technological development supporting tourism. The most recent development is the emergence of space tourism. This concept is innovatively pioneered by a number of huge technology companies like Virgin Galactic, SpaceX, Blue Origin, Orion Span, and Boeing. Even though this new tourism concept is still in its infancy, some of those companies are already developing and successfully testing their prototype of sub-orbital spacecraft. The growing interest in and popularity of space tourism will drive beyond earth orbit travel especially when space travel becomes financially and logistically feasible. Many new and more sophisticated technologies will be adopted as part of this project including the extensive use of big data analytics.

In a similar way, the marketing approach for the tourism industry is also facing big changes. Tourism companies are now moving away from conventional marketing to mass e-marketing in order to reach wider audiences. The use of "electronic word of mouth" (eWOM) through social media platforms is gaining in popularity. With the machine learning and artificial intelligence capabilities to predict the behaviour of potential customers through social media-generated big data, tourism marketers can bring tourism products beyond geographical boundaries in real-time through eWOM marketing strategies (Miah et al., 2017). eWOM serves dual purposes in tourism marketing as a source of information for tourism destinations and as a recommendation for a particular tourism destination (Kanje et al., 2020).

Another emerging technology in tourism marketing is the use of neuro-marketing technology. Neuro-marketing researchers argue that buying decisions of consumers can be influenced by things that are so subtle that even the buyers do not even notice them, such as emotion, intuition, and the unconscious mind. In the tourism context, strong human emotion can affect the way the tourists choose their travelling destination since it involves experience, feeling, and memory. As illustrated in Figure 5.1, a bad experience of a tourist will lead to his/her bad feeling which also turns into bad memory. This negative emotion scenario will result in a "no repeat" travelling decision. On the other hand, any good tourism experience triggers good feelings which turn into a very sweet memory. The final result of this positive emotion is a strong emotional connection to the tourism experience and destination which in the end turns into a "repeat travelling" decision.

Tourism marketers can use neuro-marketing technology to generate big data analysis on neurological and biological patterns in order to predict customers' attitude and reaction towards tourism products using neurological data collection techniques such as eye tracking, facial recognition systems, galvanic skin response, and heart rate (Boz et al., 2017). Figure 5.2 illustrates an example of how a neuro-marketing technique was applied when the consumers browsing through the information on a tourism website.

In addition, the hospitality industry is also competitive in capitalizing on advanced artificial technology based on big data in numerous ways. For example,

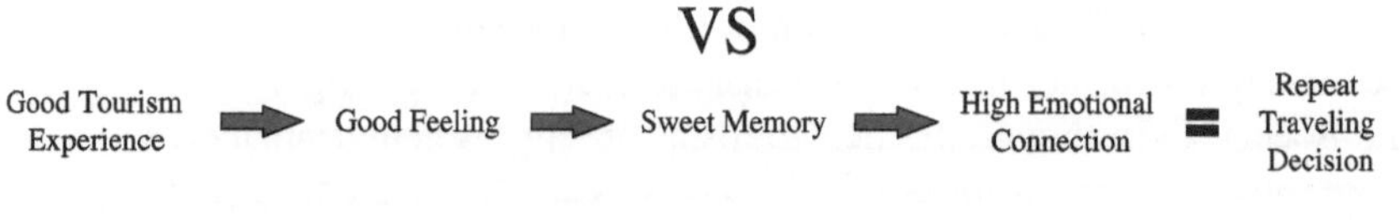

Figure 5.1 Relationship between experience, feeling, memory, emotion, and travelling decision in neuro-marketing (source: developed by the authors).

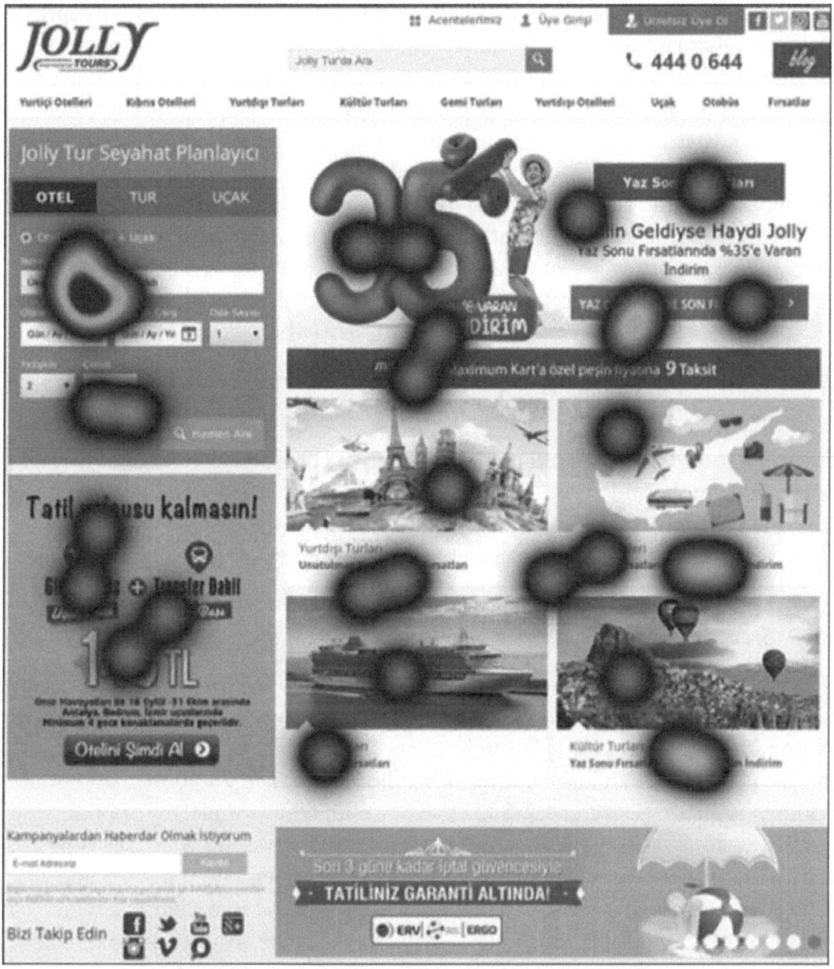

Figure 5.2 Example of neuro-marketing analysis technique applied in a tourism website (source: Boz et al., 2017).

autonomous human-robots as front desk personnel, concierge, room service assistants, butlers, and luggage porters, and the most popular one, the chatbot have been operating in a number of major hotels. The advantages of using robots in hospitality are that they do not get tired, bored, or emotionally disturbed, and most importantly they do not require any salary and are able to work 24/7. In addition, robot can overcome human errors and perform with a higher degree of consistency and accuracy than human hotel employees. On the other hand, the disadvantages of using robots in hospitality are the employment crisis due to those robots replacing human staff, the high maintenance cost, and the inability to perform work beyond their programmed parameters.

Another big data-based new technology in the modern hospitality industry is the use of virtual reality technology. Virtual reality is being used to improve customers' virtual experience by simulating the real-life situations and contexts of the hotel and services for its potential customers prior to their booking decision, making it a powerful tool to meet customer needs (Pestek and Sarvan, 2021). One of the virtual-augmented reality technology examples used is Virtual Travel Experiences (VTE). VTEs allow potential customers to virtually visit rooms and services in the hotel as well as preview local tourism attractions through 360-degree augmented reality video technology before considering the hotel's offer. The database will record the customer's virtual experience data for later usage on the next prospective customers through artificial intelligence systems.

Issues in Implementing Big Data Analytics in Aviation, Tourism, and Hospitality

The trend towards implementing big data analytics technology seems unavoidable in the present boundaryless business world. Nevertheless, while adopting big data analytics technology is considered beneficial to the aviation, tourism, and hospitality service industries, investing in this latest technology poses some issues.

Data Security and Cyber Threats

Cyber security has been a major concern in the aviation industry especially ever since the smart airport and fly-by-wire technology in the new generation of aircraft was introduced (Ukwandu et al., 2021). The busiest airports such as Chicago O'Hare International Airport handled more than 900,000 take-offs and landings in 2019 (Dave et al., 2022). As a result, the complexity of preserving cyber security at airports is also multiplied by the increase in air traffic over the years. The huge volume of data and information on aircraft and air traffic that airlines and airports need to manage may attract possible hacking activities by terrorists or irresponsible individuals. Maintaining cyber security in aviation also poses additional challenges since it requires heavy investment in technology, funds, and skilled staff (Kagalwalla and Churi, 2019).

Meanwhile, in the tourism and hospitality industries, data security has also become a pertinent issue since both are known as industries with large data pools (Yağcı et al., 2020). The tourism industry itself has been ranked as the third-highest after retail businesses and the financial sector with regard to its vulnerability to cybercrimes (Preko, 2020). The nature of tourism and hospitality businesses which requires them to use technology to collect and retain a high volume of guest information can potentially attract security threats such as fraud, hacks, and espionage. The data breach incident involving the Marriott International hotel chain is a popular example. The breach began in 2014, but it was only discovered later in 2018 that it affected nearly 500 million guests worldwide, with confidential personal data, such as contact numbers, passport numbers, and credit card numbers, of 100 million guests compromised (Mest, 2019).

Data Sharing Trust

The privacy and security of customers' data have been of major concern in the aviation, tourism, and hospitality industries (Yallop and Seraphin, 2020). One of the ways to secure customers' trust in data security and privacy is by ensuring that those data are safely and securely kept in a well-designed data governance system framework. However, to manage customers' experiences throughout their travel, the most effective method is to allow these data to be shared among multiple stakeholders across the aviation, tourism, and hospitality industries. This triggers serious trust concerns, especially from the customers' side as they worry that their personal information is being shared with and manipulated by third parties beyond their knowledge, need, and consent.

Apart from that, there are also trust concerns that come along with the use of some of the latest technology within the aviation, tourism, and hospitality industries. While big data is used to gather the real-time data of millions of end-users, it also enables access to various personal data which could either deliberately or unintentionally compromise the privacy of individuals (Nair, 2020). The use of cookie files to accumulate customers' information while interacting with a tourism company's or an airline's website or app, for example, is believed by the public to collect customers' personal data without their knowing (Sharakhina and Skvortsova, 2019), and then expose them to possible exploitation for commercial advantage (Wigan and Clarke, 2013).

Mistakes in Big Data Analytics Interpretations

The collected big data by itself does not speak for itself unless it is put into the business context. Thus, the interpretation of big data is one of the crucial processes in big data handling. However, Chareyron, Da-Rugna, and Raimbault (2014) asserted that the most challenging problem in handling big data lies in how to present the collected big data so that it can convey a piece of meaningful and understandable information to business decision-makers. Since the information is massive, very often the handlers of that information, especially managers, do not understand how and which data should be extracted, stored, and analyzed, and which should be ignored throughout their decision-making process (Ogbeide et al., 2020). In addition, the interpretation of big data is heavily subjective and prone to humans' limited cognitive capability and subjective feelings (Sharakhina and Skvortsova, 2019). Consequently, an incorrect interpretation of the meaning of big data tends to lead to wrong or inaccurate business and operational decisions. To make matters worse, in the context of aviation especially, a slight mistake in an operational decision may cause the loss of millions of dollars' worth of aircraft or even the loss of human lives.

Negative Return-on-Investment in Big Data Analytics Technology

Big data analytics technology demands a very high investment from companies. While big data analytics technology offers many benefits to aviation, tourism, and hospitality businesses, some companies, especially the small- and medium-size companies, perceive that investment in big data analytics technology does not give them value for their money. As highlighted by Ghafar, Harun, Lailawati, and Rahman (2020), there are two main factors that make some small- and medium-sized companies reluctant to invest in Industry 4.0-related technology including big data analytics. First, they perceive that initial investment in big data analytics technology as well as sustaining it requires extensive monetary funding, which is beyond what they can afford. Second, they also worry that when they invest in giving expensive training on big data analytics to their employees, there is still no guarantee that those trained highly skilled employees will remain loyal to the companies, as big data analytics skills are extremely in demand with competitors. As a result, many perceive the investment in big data analytics technology as a negative return-on-investment business move.

Big Data Analytics in Unprecedented Air Travelling Environment

History has evidenced that the aviation, tourism, and hospitality industries' activities are easily caught up with the occurrence of uncertainty, shocks, and unprecedented global events. The Gulf War in 1991, the financial crisis in 1997, the 9/11 attacks, the severe acute respiratory syndrome (SARs) outbreak in 2003, and the global economic crisis in 2008 are among the unprecedented events that have badly affected the aviation, tourism, and hospitality industries in the past (Sobieralski, 2020), causing many travelling bans and restrictions to be imposed in many countries.

Nevertheless, the present outbreak of the COVID-19 pandemic is the worst unprecedented event of this new millennium era. The present pandemic that has caused global air travel restrictions shows how unprecedented situations and lack of knowledge to plan for and manage risks can put the aviation, tourism, and hospitality industries in grave peril (Gallego and Font, 2021). Without a doubt, the recent pandemic has created upheaval around the world by reshaping the travelling activities and behaviours as well as mobility styles and habits of all people around the world (Shamshiripour et al., 2020). As a result of the complex implications of the pandemic, big data analytics technology is being employed to eliminate or minimize the infection risks of the deadly virus without forbidding people from enjoying their travelling and tourism activities.

However, other unprecedented events of a different nature may happen in the future. The recent COVID-19 pandemic raises the question of sustainability in the industries. The pandemic has taught global leaders and industry experts that regardless of how advanced technology is, it still cannot predict some unprecedented events and the intensity or the magnitude of their impacts and disruptiveness for aviation, tourism, and hospitality business activities.

Conclusion

Big data is considered the new "fuel" for many technologies including machine learning, artificial intelligence, virtual reality, and augmented reality applications.

Since the aviation, tourism, and hospitality services are generating massive data every second on a daily basis throughout travellers' journeys, tourism experiences, and hospitality services experiences, reliance upon this new informational resource capital is inevitable. Although these technologies have already penetrated the aviation, tourism, and hospitality industries in most Southeast Asian countries, their adoption is still in its infancy. There are many opportunities for the aviation, tourism, and hospitality operators to explore and invest in big data analytics technology (Yallop and Seraphin, 2020), and the most important point is that the need and opportunities for big data analytics in these industries will continuously increase in the future. However, there are also some issues or challenges that industry leaders and experts need to address with practical and strategic business solutions so that the growth of the aviation, tourism, and hospitality industries is continuously sustained.

References

Ardito, L., Cerchione, R., Del Vecchio, P. and Raguseo, E. (2019). Big data in smart tourism: Challenges, issues and opportunities. *Current Issues Tourism*, 22(15), pp. 1805–1809.

Boz, H., Arslan, A. and Koc, E. (2017). Neuromarketing aspect of tounsm pricing psychology. *Tourism Management Perspectives*, 23, pp. 119–128.

Brar, K. (2019). Hotels using big data to check out guests. *SPH Digital News*. Retrieved from: https://www.tnp.sg/news/views/hotels-using-big-data-check-out-guests (accessed: 19th August, 2021).

Chareyron, G., Da-Rugna, J. and Raimbault, T. (2014). Big data: A new challenge for tourism. In 2014 IEEE International Conference on Big Data (Big Data) (pp. 5–7). Washington, DC, USA: IEEE. The 27th October-30th October, 2014.

Dave, G., Choudhary, G., Sihag, V., You, I. and Choo, K. K. R. (2022). Cyber security challenges in aviation communication, navigation, and surveillance. *Computers & Security*, 112, p. 102516.

Gallego, I. and Font, X. (2021). Changes in air passenger demand as a result of the COVID-19 crisis: Using big data to inform tourism policy. *Journal of Sustainable Tourism*, 29(9), pp. 1470–1489.

Ghafar, N. H., Harun, K., Lailawati, M. S. and Rahman, N. A. A. (2020). Preliminary investigation of industry 4.0 readiness among small and medium aviation companies in Malaysia. *Journal of Physics: Conference Series*, 1573(1), p. 012030.

GovInsider (2018). *How the Singapore Tourism Board Uses Data to Personalise Tourism?* Retrieved from: https://govinsider.asia/smart-gov/how-the-singapore-tourism-board-uses-data-to-personalise-tourism/ (accessed: 10th August, 2021).

Hausladen, I. and Schosser, M. (2020). Towards a maturity model for big data analytics in airline network planning. *Journal of Air Transport Management*, 82, p. 101721.

Kagalwalla, N. and Churi, P. P. (2019). Cybersecurity in aviation: An intrinsic review. In 2019 5th International Conference On Computing, Communication, Control and Automation (ICCUBEA) (pp. 1–6). Pune, India: IEEE. The 19th–21st September, 2019.

Kanje, P., Charles, G., Tumsifu, E., Mossberg, L. and Andersson, T. (2020). Customer engagement and eWOM in tourism. *Journal of Hospitality and Tourism Insights*, 3(3), pp. 273–289.

Mest, E. (2019). *Marriott Ends 2018 Ahead Despite Data Breach, Strikes.* Retrieved from: http://hdl.handle.net/10919/88403 (accessed: 11th August, 2021).

Miah, S. J., Vu, H. Q., Gammack, J. and McGrath, M. (2017). A big data analytics method for tourist behaviour analysis. *Information & Management*, 54(6), pp. 771–785.

Nair, S. R. (2020). A review on ethical concerns in big data management. *International Journal of Big Data Management*, 1(1), pp. 8–25.

Ogbeide, G. C., Fu, Y. Y. and Cecil, A. K. (2020). Are hospitality/tourism curricula ready for big data?. *Journal of Hospitality and Tourism Technology*, 12(1), pp. 112–123.

Pestek, A. and Sarvan, M. (2021). Virtual reality and modern tourism. *Journal of Tourism Futures*, 7(2), pp. 245–250.

Pranita, D. (2018). Digitalization: The way to tourism destination's competitive advantage (case study of Indonesia marine tourism). *KnE Social Sciences*, 3(11), pp. 243–253.

Preko, A. (2020). Tourism development: National policies and tourism priorities in Ghana. *International Journal of Tourism Policy*, 10(4), pp. 380–391.

Shamshiripour, A., Rahimi, E., Shabanpour, R. and Mohammadian, A. K. (2020). How is COVID-19 reshaping activity-travel behavior? Evidence from a comprehensive survey in Chicago. *Transportation Research Interdisciplinary Perspectives*, 7, p. 100216.

Sharakhina, L. V. and Skvortsova, V. (2019). Big data, smart data in effective communication strategies development. In 2019 Communication Strategies in Digital Society Workshop (ComSDS) (pp. 7–10). IEEE (accessed: 10th April, 2019).

Sobieralski, J. B. (2020). COVID-19 and airline employment: Insights from historical uncertainty shocks to the industry. *Transportation Research Interdisciplinary Perspectives*, 5, p. 100123.

Sritama, S. (2018). TAT eyes big data platform. *Bangkok Post.* Retrieved from: https://www.bangkokpost.com/business/1560542/tat-eyes-big-data-platform (accessed: 10th August, 2021).

Tourism Authority of Thailand (TAT) (2019). TAT announces big data partnership with grab. *TAT News.* Retrieved from: https://www.tatnews.org/2019/02/tat-announces-big-data-partnership-with-grab/ (accessed: 17th August, 2021).

Ukwandu, E., Farah, M. A. B., Hindy, H., Bures, M., Atkinson, R., Tachtatzis, C. and Bellekens, X. (2021). *Cyber-Security Challenges in Aviation Industry: A Review of Current and Future Trends.* Retrieved from: https://arxiv.org/pdf/2107.04910 (accessed: 10th August, 2021).

Vajirakachorn, T. and Chongwatpol, J. (2017). Application of business intelligence in the tourism industry: A case study of a local food festival in Thailand. *Tourism Management Perspectives*, 23, pp. 75–86.

Wee, T., Perdana, A. and Remy, D. (2019). Data analytics dilemma at Alpen Hotel. *Journal of Information Technology Teaching Cases*, 9(2), pp. 58–63.

Wigan, M. R. and Clarke, R. (2013). Big data's big unintended consequences. *Computer*, 46(6), pp. 46–53.

World Tourism Organization (UNWTO) (2020). World tourism barometer. *Statistical Annex*. 18(5). Retrieved from: https://www.e-unwto.org/doi/pdf/10.18111/wtobarometereng. 2020.18.1.5 (accessed: 18th August, 2021).

Yağcı, K., Akçay, S., Efendi, M. and Öztürk, H. M. (2020). Information security awareness in tourism enterprises: Case of Turkish manager opinions. In A. Akbaba and V. Altıntaş (Eds.), *Industrial and Managerial Solutions for Tourism Enterprises*. Hershey: IGI Global, pp. 251–267.

Yallop, A. and Seraphin, H. (2020). Big data and analytics in tourism and hospitality: Opportunities and risks. *Journal of Tourism Futures*, 6(3), pp. 257–262.

Yuniarni, S. (2017). *Indonesian Tourism Goes Digital, Partners with Thailand*. Jakarta Globe. Retrieved from: https://jakartaglobe.id/lifestyle/tourism-goes-digital/ (accessed: 10th August, 2021).

6 Technology and In-Flight Catering

How Do They Relate to Air Travel, Tourism, and Hospitality?

Fathien Azuien Yusriza and Nor Aida Abdul Rahman

Introduction

Global technology spending in the year 2022 is expected to reach US$5.3 trillion and expected to increase 5% on a compound annual growth rate (CAGR) by the year 2024. On the other hand, the global in-flight catering market or airline catering business was valued at US$18,081 million in the year 2018, and by the year 2026, it is expected to increase to US$24,113 million (Mlitz, 2021). The expansion of market value indicates the progress of rapid technology innovation. The acceleration of technology has become a source of organizational innovation to achieve operational efficiency, as well as sustain in the industry.

The airline catering sector is very much connected with travel and tourism. Travellers or tourists move from one destination to another destination along their journey with multiple modes of transportation including by air. Many travellers use air transport as their main form of transportation to reach their tourism destination.

As they travel using airlines, of course the buying patterns of the airline tickets among travellers are influenced by many factors including the selection of the in-flight meals provided while on board. Normally, travellers select the airline organization depending on the hospitality services provided by the airline organization, and also their provided in-flight meals. Technological innovation and digitalization help the airline caterer to be more innovative in producing more quality foods at a larger scale, and in a shorter period.

According to Jones (2004), airline caterers are responsible for providing all meals and beverages to the airline. Additionally, they are also responsible for providing other related products such as amenity kits, paper goods, magazines, and blankets, as well as headsets. In-flight meals vary widely depending on the duration of the flight. There are specific menus for short-haul and long-haul flights; business class meals, first class meals, and economy class meals also differ between full-service airline and low-cost airlines. Normally, the production of the in-flight meals is done near the airport. This is to ensure that the foods or the in-flight meals loading onto the aircraft are appropriate and immediate. This is also important to ensure that any urgent request on in-flight meals can be fulfilled immediately.

DOI: 10.4324/9781003315179-9

Technology applications in the airline context are vital since the level of daily food production is high. With the technology applications and the appropriate system used, the alignment of the food order system and the quality as well as security of the food can be continuously measured. At the same time, with the technology or system used, the wastage of airline food meals at the kitchen and from the aircraft can be managed effectively. This chapter is crafted with three objectives. The next section will discuss the role of technology in airline catering, followed by current and future technology in airline catering organizations. The chapter then will also touch on several airline caterers as case examples of technology applications. To enhance current research on technology applications in the airline catering context, based on earlier discussion of this technology issue and practice, this chapter concludes with providing an agenda for future research opportunities.

The Role of Airline Catering and Technology in the Catering Business

In general, the role of technology in airline catering is extensive. As highlighted by a *Forbes* report by Martin (2019), technology has changed how the caterer produces food. Rapid innovation and technology have improved the process of producing in-flight meals, as well as how they are packaged. In fact, also due to technology, the food is safe to be kept for a certain duration. At the same time, the technology used in catering leads to increased quality of the food, and maintains the safety and security of the food. Information technology (IT) and the systems used in catering businesses help increase productivity, and make the integration of food reservation and food production effective.

It is acknowledged that the aviation sector consists of many players such as the airlines, cargo operators, airline caterers, airports, and many others. One of the important organizations in the aviation sector is the airline catering organization as it supplies the in-flight meals for the airlines and the passengers on board. Airline catering kitchens prepare high quantities of in-flight meals that need to be assembled. Each of these produced in-flight meals has its own characteristics such as being served as a main course, starter, snacks, or sweets. The production of the meals also depends on the schedules of the flights, weather at the destination, short-haul or long-haul journey, and characteristics of the airline such as low-cost carrier, hybrid, or full-service carrier. All of these are important factors in producing food or meals from the perspective of airline catering.

Thousands of meals are prepared every day, and all of this food is ready-to-eat meals. In fact, all of these meals are served to the travellers or passengers on board at a certain height in the air. The quality and the safety of the food are the top priority issues. For this reason, these airline catering operations are guided by various rules and regulations to ensure the safety of the passenger when they consume the food. For instance, airline caterers in Malaysia are guided by the International Air Transport Association (IATA), as well as the Ministry of Health

(MOH). Issues pertaining to food waste in the aircraft are also guided by the IATA cabin waste handbook.

Among the key roles of technology application in the catering industry are large production scale, system integration, and better efficiency in production, as well as the quality, safety, and nutritiousness of the food. Figure 6.1 and Table 6.1 highlight the key roles of technology from the perspective of airline catering.

With the growth and development of technology and the food industry, the application of technology in the food and beverage industry, including in the airline catering context, has benefitted many parties involved, especially the food producer. As highlighted above, technology helps the food producer or the food caterer to produce premium quality menus and food, as well as ensure quality and safety during consumption.

The demand for healthy food among travellers including passengers on board is increasing. As such airline caterers are responsible for ensuring their in-flight meals are nutritious and safe to be consumed. In fact, the use of technology in their in-flight meal production not only increases its efficiency, but also helps in integrating all supply chain players in the network. The digitalization and automation systems in airline catering organization could improve the company organization financially and managerially.

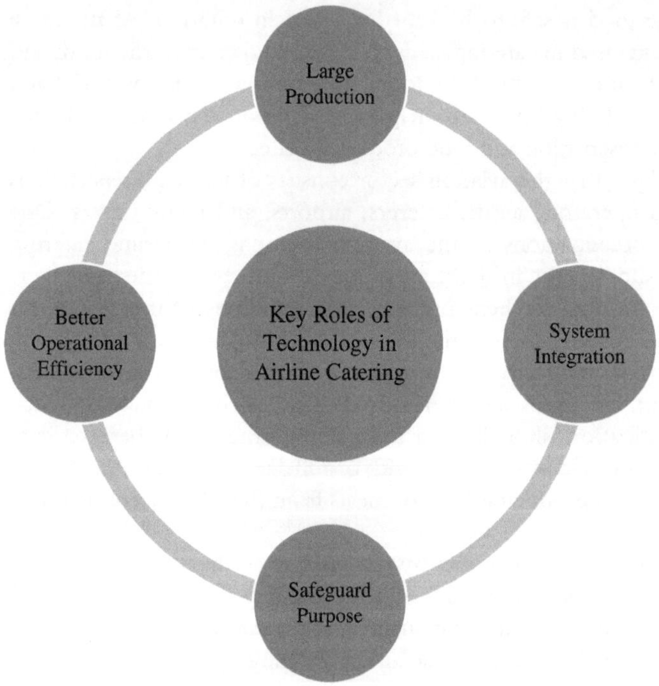

Figure 6.1 Key roles of technology in airline catering (source: developed by the authors).

Table 6.1 Key roles of technology in in-flight catering

Roles of technology	*Description*	*References*
Large production scale	Smart technology used in the food industry including in-flight catering helps in large-scale production activity in many aspects including the fermentation process. Additionally, it also helps to control waste management efforts in large-scale production activities.	Desobgo (2018); Thamagasorn and Pharino (2019)
System integration	The relevancy of technology in the food industry or catering is also recognized as it ensures the integration of the channel members channel and improves integration of the activities involving suppliers along the chain. At the same time, it helps to facilitate the information sharing process among the players in a network. Information on raw materials is easily traced without any inherent risk.	Vodenicharova (2020); Dehghani et al. (2021)
Better efficiency in production	Technology improves efficiency in food production. In fact, it can also advance the safety of the food and shelf life of the product. Additionally, apart from increasing productivity, technology also helps to cut down the cost of food storage and improve processing and packaging, and it can improve the shelf life and safety of food. The use of machines in the food industry also ensures quality and affordability.	Yang et al. (2020)
Safeguard purpose: quality, safety, and nutritiousness of food	Technology usage in the food and beverage industry including airline catering could improve the safety and quality of the food produced. It also helps to secure the nutritiousness of food produced.	Kavsara et al. (2020)

(Source: developed by the authors.)

In fact, processes like foreseeing and surveying the ideal menu and food safety could be streamlined using the appropriate system or technology. Different technology such as robotics and AI also make airline caterers and other food producer organizations more secure and quicker in process (Yang et al., 2020). With that, it is vital to note that technology promises a superior utilization of nutrients from farm to fork throughout the food production process. Technological advancement in the airline catering context is still dearth in previous study. There is a significant gap in discussion on the usage of technology applications in a specific field airline catering to get further detailed information on the benefits and challenges of technology application in the airline catering sector. Hence, it can be concluded that we need more future studies to investigate the issue empirically from many angles and perspectives. The next sub-section will briefly explain the current system and future technology adoption from the airline catering perspective.

Current System and Future Technology Use in Airline Catering

Aero Chef System in Airline Catering

The Aero Chef System is widely used in the airline catering business. It is a web-based technology application that controls every function of airline catering and works related to in-flight catering management. In general, the system integrates multiple divisions of airline catering and the operational activities at the kitchen to ensure the quality and accuracy of the meals provided. It is very important to have streamlined and transparent information among the kitchen and other departments. The information also needs to be streamlined between the airline as well as the airline catering organization (Aerochef Global, 2021). With this system, the connection and updating work on fleet information, menu planning, load planning, galley planning inventory, and asset equipment planning could be managed effectively and efficiently.

SITA System in Airline Catering

SITA is a worldwide IT provider in the aviation sector. Initially, the SITA system was also used in the airline catering sector to forecast the demand for passenger in-flight meals.

ERP System and Airline Catering

Enterprise resource planning (ERP) is the system used in an organization to control the internal processes of the organization in certain activities such as supply chain, accounting, sales, inventory management, and many more. It refers to the system for recording your business data and information. Among the advantages of using this technology in the airline catering business are as illustrated in Figure 6.2.

Radio-Frequency Identification Technology and Airline Catering

It is acknowledged that radio-frequency identification (RFID) technology allows the process at the kitchen to be fully optimized. In air travel, from the perspective of the airline, the precise tracking of the passenger in-flight meals is needed, and

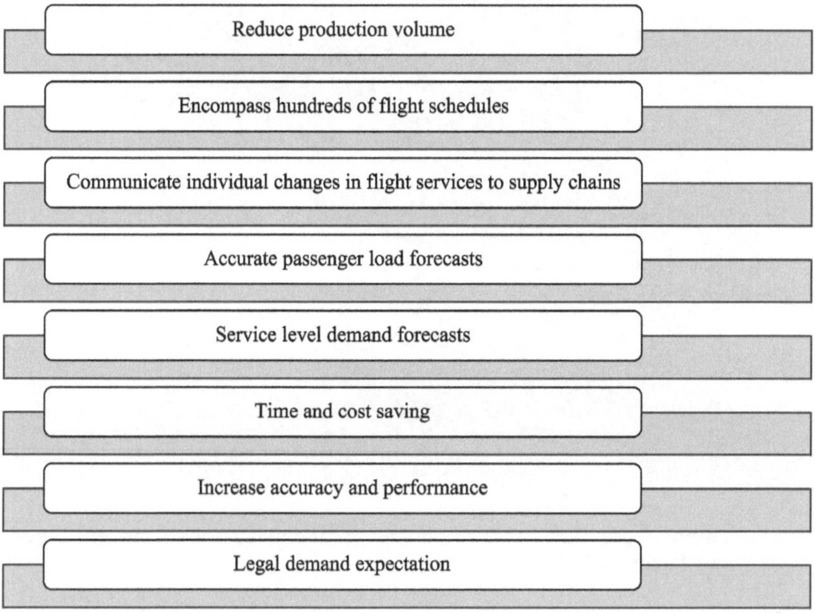

Figure 6.2 The advantages of using the ERP system in the airline catering sector (source: developed by the authors).

it should be connected directly to the type of aircraft that carries the travellers. For instance, this latest technology is used in the current operation of the Airbus A380 Super Jumbo. RFID technology is used and connected to this type of aircraft to distribute the meals according to an exact planned timeline. As reported in Pepperl and Fuchs (2021), RFID technology enables the exact correlation between the traveller (passenger) and their in-flight meals. For airlines using latest design of the Airbus aircraft A380. It can benefit the culinary amenities that is controlled by RFID.

Additionally, RFID tagging technology is also used in the catering industry. In the context of in-flight catering, this technology is largely used as it can be attached to the catering containers and trays. The data is accessible during the heating process with induction ovens.

As for the travellers, one of the benefits to travellers or passengers onboard is that they can enjoy the hot meal on the airplane and feel at home. It improves the customer experience. In fact, this type of technology can reduce the numbers of ovens on board. And as for the advantages for the airline, they can optimize the space and offer the space as a paying passenger seat. To conclude, with this RFID technology, the airline can design their aircraft configuration efficiently and ensure the space is fully maximized to the fullest for every purpose. What is more, this RFID technology at present could improve time efficiency as well, since the trolleys could be equipped with this technology. With that, trolleys

Figure 6.3 Onboard tray with RFID technology for individual trays (source: Pepperl and Fuchs, 2021).

could be easily identified and sent to the correct aircraft and airline. Figure 6.3 shows an onboard tray with RFID technology for individual trays.

Technology and Airline Catering: Some Cases and Examples

KLM Airlines and Robotic Technology in Airline Catering

It was reported by Hospitality (2020) that KLM Royal Dutch Airlines has adopted robotic technology in producing in-flight meals for their economy class travellers/passengers. The use of robotic technology could reduce handling costs at the kitchen during production, and at the same time, it can increase the flexibility and passenger centricity. In fact, this technology was taken into production in the airline catering business after the COVID-19 outbreak. It was reported also that, at present, only this airline catering organization is utilizing this robotic technology.

Lufthansa and Smart Catering Technology

In the case of Lufthansa catering, technology used such as big data application helps the caterer get more information on their passenger or customer feedback. From the feedback, it develops new in-flight meals and menus. Big data and data analytics technology helps the industry to filter the information gathered from passengers or customers, and helps them in improving their menus and catering concept.

As reported by Lufthansa Industry Solutions (2021), technology use in airline catering also assists the airline caterer to plan the raw material requirements more efficiently. Precise material planning in in-flight meals production could help the airline caterer to achieve cost optimization and reduce waste. In fact, with appropriate system integration technology, the airline caterer could cope with a growing number of urgent demands and special menu demands. Additionally, digitalization allows the airline caterer to be more flexible and ready to adjust their scheduled programme and reduce the manual effort.

Etihad, Artificial Intelligence Technology and Food Waste Reduction Efforts

In Etihad, for instance, AI technology was used to reduce food wastage in airline catering. The AI technology application in the national airline of the United Arab Emirates has reportedly partnered with Lumitics company, a Singaporean food technology organization. They have been working together in the effort to reduce food wastage both in the kitchen and in the aircraft. The use of machine learning (ML) technology, a popular domain of AI technology, matched with computer vision usage, helps the industry to identify wastage patterns across the airline supply chain network (Etihad, 2020). It is also able to track the unconsumed in-flight meals in the airplane from economy class. With this information, the airline caterer could use the information to improve their meal planning and achieve cost reduction.

Conclusion, Limitations, and Future Research Opportunities in Airline Catering and Technology

Despite the fact that this chapter offers valuable discussion on the sector of airline catering and technology, the main limitation of this study is that it is not empirically tested or explored. The discussion provided in this chapter uses an abductive approach hence limit the discussion more detail. The awareness of the technology applications delivered in this chapter, focusing on the roles and types of technology, allows the reader to have a concise understanding. Future scholars are called to further investigate this issue empirically for a further and deeper understanding of how the extent of technology application in the airline catering context could overcome operational and decisional challenges in improving performance and achieving sustainability. It is therefore proposed that future studies also investigate the role of governments and other stakeholders involved to build and improve current technology practice in the airline catering context, especially in developing countries. This is important for achieving sustainable and successful technology application in the industry.

References

Aerochef Global (2021). *For Airlines, Integrated Software Solution for Airlines with Catering Management.* Retrieved from: https://www.aerochefglobal.com/airlines.aspx (accessed: 3rd September, 2021).

Dehghani, M., Popova, A. and Gheitanchi, S. (2021). Factors impacting digital transformations of the food industry by adoption of blockchain technology. *Journal of Business & Industrial Marketing*, https://doi.org/10.1108/JBIM-12-2020-0540

Desobgo, S. C. Z. (2018). Modernization of fermenters for large-scale production in the food and beverage industry. In S. Panda and P. Shetty (Eds.), *Innovations in Technologies for Fermented Food and Beverage Industries. Food Microbiology and Food Safety.* Cham, Springer: pp. 189–220.

Etihad (2020). *Etihad Airways Teams up with Lumitics to Reduce Inflight Food Wastage.* Retrieved from: https://www.etihad.com/en/news/etihad-airways-teams-up-with-lumitics-to-reduce-inflight-food-wastage (accessed: 27th November, 2021).

Hospitality (2020). KLM adopts innovative robot for inflight catering production. *Future Travel Experience.* Retrieved from: https://www.futuretravelexperience.com/2020/10/klm-adopts-innovative-robot-for-inflight-catering-production/ (accessed: 22th August, 2021).

Jones, P. (2004). *Fligh Catering.* Abingdon: Routledge.

Kavsara, H. K., Ozilgen, S. and Dagdeviren, M. (2020). Safeguarding grandma's fermented beverage recipes for food security: Food safety challenges. *International Journal of Gastronomy and Food Science*, https://doi.org/ 10.1016/j.ijgfs.2020.100266

Lufthansa Industry Solution (2021). *Smart Catering Serves up Food Trends.* Retrieved from: https://www.lufthansa-industry-solutions.com/de-en/solutions-products/system-catering-and-airline-catering/smart-catering-serves-up-food-trends (accessed: 21st November, 2021).

Martin, N. (2019). *How Technology is Transforming the Food Industry.* Forbes. Retrieved from: https://www.forbes.com/sites/nicolemartin1/2019/04/29/how-technology-is-transforming-the-food-industry/?sh=770a0edd20a3 (accessed: 2nd January, 2022).

Mlitz, K. (2021). Global information technology industry forecast 2019–2022, by region. Statista. Retrieved from: https://www.statista.com/statistics/507365/worldwide-information-technology-industry-by-region/ (accessed: 12th December 2021).

Pepperl and Fuchs (2021). *RFID Enables Catering.* Retrieved from: https://www.pepperl-fuchs.com/global/en/20057.htm (accessed: 25th December, 2021).

Thamagasorn, M. and Pharino, C. (2019). An analysis of food waste from a flight catering business for sustainable food waste management: A case study of Halal food production process. *Journal of Cleaner Production*, 228, pp. 845–855.

Vodenicharova, M. S. (2020). Supply chain study in food industry in Bulgaria. *International Journal of Retail & Distribution Management*, 48(9), pp. 921–938.

Yang, L., Ma, C., Yang, Y., Zhang, E. and Lv, H. (2020). Estimating the regional eco-efficiency in China based on bootstrapping by-production technologies. *Journal of Cleaner Production*, 243, p. 118550.

Part 3

The Southeast Asia Specific Cases

7 Digital Transformation in the Thailand and Malaysian Tourism and Aviation Industries

Suhaila Abdul Kadir and
Siti Zaleha Noraini Binti Hussin

Introduction

Across industries, technology is evolving at incredible speeds. The ability of a country to be well governed is strongly influenced by the rapid growth of technology (Khairul Nizam, 2006). As a result, industrial companies must be adventurous in their application of technology to promote products to global consumers (Jawing, 2020). Tourism, hospitality, and aviation are among the industries transforming their usage of technological applications (Alrashid, 2012). With the spread of digital technologies, the tourism and aviation industries are likewise growing at a faster pace. The tourism and hospitality industry's future has been shaped by current developments in Internet of things (IoT) technology (Kansakar et al., 2018). Handset Profile (HSP) has to be on the cutting edge of IoT technology to maintain a competitive edge in the business. Sensors, actuators, identification tags, mobile devices, and other similar devices that may communicate directly or indirectly with one another via local communication networks or the Internet are the linking of ordinary physical items (Munir et al., 2017).

Digital Transformations in the Aviation Industry

In response to the expansion of digital innovation, AirAsia has used face recognition-based security inspection equipment in the departures area of Avalon International Airport. The first cross-border route to offer biometric facial recognition services was the Fast Airport Clearance Experience System (FACES), which is owned and operated by a low-cost carrier (Derus, 2019). In Southeast Asia, FACES is the first self-check-in system to employ face recognition software. After checking in, passengers may board the aircraft without presenting any travel documents. FACES adoption also enhances the effectiveness and speed of operations at Avalon International Airport.

Thailand has one of the most competitive aviation markets in the world. Because of this, customer satisfaction with the level of airport service and the effective management of its facilities is a critical element that establishes market distinctiveness and, consequently, influences the long-term survival of the business. To increase commercial revenue and establish itself as a hub in Southeast Asia, the

DOI: 10.4324/9781003315179-11

airport must launch a new era of technological innovation that handles the whole IT infrastructure of the facility within the context of a service that prioritises high-quality services for airport consumers. The information technology (IT) system and application must be connected to the community's tourism sector for attraction and communication with associated sectors (Charernnit, 2020).

Robotics

Due to the quick development of robotics, automation is becoming more popular in the aviation sector. An autonomous luggage transporter, the latest Fraport robot YAPE. Particularly at Incheon Airport, YAPE is a self-driving luggage mover. Just a few current examples include British Airways' staff working with a startup company "BotsAndUs" at Heathrow Terminal 5 to test AI-powered autonomous robots to increase passenger punctuality, the robot named "Airstar" at Incheon Airport, and Josie Pepper at Munich Airport.

Thailand is using robotic technology in their airports; Changi is cleaned by a number of robots, and Jewel Changi Airport (Jewel) has a variety of robotic assistance. Each one has a specific role to play in maintaining Jewel's buildings, doing labour-intensive work, or offering extra-special services. The robots are categorized as mop-sweeping robots at Level 1 of Jewel, near the Rain Vortex. Guests can also communicate with the robots by hitting the button in the centre of the robot's chest, which allows them to scan a QR code. Aside from that, there are robots that do the heavy lifting and those that provide a touch-based service. The robots that perform touch-based service at Jewel's Canopy Park (on Level 5) emerge, providing candy and bottled water to guests who have spent their energy sliding, bouncing, or exploring the variety of recreational activities (Changi Airport Group, 2021). Furthermore, with the recent advancements in drone technology, Edmonton International Airport (EIA) recently established a joint venture with Drone Delivery Canada (DDC) to work together to develop the airport into a hub for drone cargo transportation in Western and Northern Canada.

Digital Twin

The operations section has been shown a fully functional digital twin from the SITA Lab that is being tested at an airport on the East Coast of the United States using the 3D interface of an 86-inch touch-screen centre. The digital twin will enhance decision-making since it provides a complete picture of airport operations (Future Travel Experience, 2020). Data flows need to be the same for the digital twin idea to function. A real-time transportation system is used to verify the information obtained from an airport. The machine learning training for the centralized control system requires this synthetic data. A knowledge-based model that uses data streams that are updated in-flight in real time is the true definition of a digital twin. In other words, the idea was inspired by the Internet of Things (IoT) since a digital twin bridges the virtual and real worlds (Saifutdinov et al., 2020).

Over the past few years, the aviation industry has shown a strong commitment to realizing the potential of Artificial Intelligence (AI) through a variety of applications. On the one hand, chatbots are now being used by airlines and airports to interact with passengers and enhance operations. At the 2019 FTE APEX Asia EXPO Awards, AirAsia created and utilized AirAsia Virtual Allstar (AVA), which was recognized with a Silver Award for Best Passenger Experience Initiative in the Airline Category. Real-time updates reduce the impact of interruptions, luggage delays are decreased, and individualized information is provided to customers via digital channels, all of which are advantages for passengers (Future Travel Experience, 2020).

(5G Networking and Services)

Thailand's aviation agencies signed a 5G Smart Airport Memorandum of Understanding (MoU) in 2020 to establish 5G networks and services at the nation's airports. In accordance with the conditions of the MoU, the 28 other airports managed by the Department of Airports (DoA) across the nation will receive 5G technology solutions when they have been created and implemented at Krabi Airport. The advancements represent the first national-scale 5G smart airport initiative in Southeast Asia. GlobalData predicted the Asia Pacific (APAC) region will lead the world in 5G apps and usage by 2024, projecting 1.14 billion APAC consumers would account for roughly 65% of global 5G subscriptions within the next four years before the coronavirus pandemic swept the globe. (Devanesan, 2020).

Inflight Connectivity

With an increasing number of airlines attempting to digitalize their inflight experience, the future of the inflight connectivity sector is bright. This demand is generating a significant improvement in the degree of connection available (Future Travel Experience, 2020). Airlines such as Bangkok Airways, Thai AirAsia, and Thai Lion Air (TLA) are not hurrying to give cloud access due to high fees and surfing speeds that aren't always up to the demands of the tech-savvy tourist. As a result, just two Thai airlines, the flag carrier Thai Airways International (THAI) and Nok Air, a low-cost flight, have launched inflight broadband Wi-Fi (Kositchotethana, 2014).

Biometrics Technology

According to a recent report, "Future Travel Experience, 2020", in 2020, more airlines are anticipated to release smartphone apps for facial recognition, improving the convenience of the check-in process for customers. Iberia is testing a facial recognition app at Madrid Airport, which will enable customers to identify themselves using their biometric profile at both standard and fast-track security checkpoints, as well as at boarding gates, removing the need to present travel documents. Furthermore, because biometric authentication has advantages and disadvantages, deploying a single biometric authentication approach might not be appropriate. In order to identify biometric features, each biometric sensor may employ a different method. Additionally, to assist in boosting security, researchers suggest employing a number of biometric authentication techniques. The accuracy of the procedure was 99.82% when using biometric authentication to confirm the face and hand (Pukdesree and Netinant, 2016).

Thailand erected 2,000 biometric checks at all border crossings in May 2019, in contrast. This summer, the Thai Ministry of Foreign Affairs (MOFA) contracted a group led by Gemalto to produce 15 million biometric electronic passports by 2026. (Pascu, 2019). The Personal Data Protection Act B.E.2562 (PDPA), Thailand's first data protection law, is one of many pieces of Thai legislation that mentions guidelines for the security of biometric data for the acquisition, use, and disclosure of personal data, including biometric data, and defines a fundamental set of rules. Studying the General Data Protection Regulation of the EU (GDPR), which the PDPA generally follows, the UK Data Protection Act 2018 and the UK's Information Commissioner's Office (to learn how the member states implement the GDPR's rules), and the Illinois Biometric Information Privacy Act (the first biometric data protection act of the USA), would be the best examples for Thailand to amend the Act and have a guideline for processing bio (Chuenpukdee, 2020).

Sustainability

International demand for air travel is predicted to grow by an average of 5–6% annually over the long run. Both positive and negative effects can result from this study. Aviation has a significant cultural impact on us. Airlines and airports are looking towards collaborating with younger, more adaptable enterprises as a result of the technical developments discussed in the paragraphs above to help re-imagine the journey. Many people are now able to go to several locations worldwide in a single day because of the sharp decline in ticket prices. Some contend that unfavourable outcomes, such as noise pollution and greenhouse gas emissions, won't get worse in the future because strategies to control them are being developed (De Haan, 2010).

Assistive Tech: Self-Driving Electric Wheelchairs and Accessible IFE

The way airports and airlines cater to passengers with special needs is changing, due in part to technological improvements. As part of a comprehensive effort to expand mobility and accessibility services at Tokyo Narita International Airport in 2020, All Nippon Airways (ANA) has partnered with Panasonic Corporation to test the newest generation of personal mobility self-driving electric wheelchairs. While at Abu Dhabi International Airport, Etihad Airways and Abu Dhabi Airports completed guest trials of innovative autonomous wheelchairs in December 2020. Over 60 visitors with limited mobility participated in the testing; they self-drove the wheelchairs through congested areas and lounges before making it safely to their gate (Future Travel Experience, 2020).

Hearable, Wearable, and Voice Technology

Despite the fact that Google Glass may not have met expectations, the airline sector continues to experiment with wearable solutions. For example, Spryng, a smart device developed to assist travellers in feeling more refreshed while en route, is being offered by Sri Lankan Airlines in collaboration with MAS Holdings. Spryng is a wireless, dynamic pneumatic compression wrap that helps to alleviate muscle soreness by enhancing blood circulation and increasing cellular oxygen delivery (Future Travel Experience, 2020).

eVTOL and Autonomous Travel

In order to encourage the growth of commercial air taxi services, a number of projects were implemented in 2019. In October, Boeing, Bell, Embraer, Safran, Uber, Fraport, Groupe ADP, and other major corporations completed the first manned flight over Singapore, while other major corporations such as Boeing, Bell, Embraer, Safran, Uber, Fraport, Groupe ADP, and others announced plans for urban air mobility solutions. Uber Air is testing its flying taxi services all around the world; thus flying automobiles are becoming a reality as well. Uber Air trials will begin next year in Melbourne, with commercial operations set to begin in 2023, according to the firm (Future Travel Experience, 2020).

Cooperation with Startups

As a result of the technological advancements described in the preceding paragraphs, airlines and airports are looking into forming collaborations with younger, more dynamic businesses to assist in re-imagining the journey itself (Future Travel Experience, 2020). Air Lanna, a projected Thai full-service airline founded by Bangkok-based CEI Aviation and Logistics Co., Ltd, proposes to provide scheduled and charter flights from Bangkok to local locations, as well as to China, Southeast Asia, and the Middle East, with an initial fleet of two B737-800s (AVIATOR, 2018).

Digital Transformation and Tourism

The terms "tourism" and "travel" have distinct meanings. Tourism, according to the Hall's Dictionary, comes from the word "lancong," which implies visiting while looking around (having fun) (Noresah, 1997). While travelling can be defined as travel or journey in English (Hawkins, 2006). Travelling can also be defined as leaving home for a specific purpose and planning to return to the starting point (Bahrom et al., 2010). T

Tourism is a key service industry that is expanding and generating significant revenues in the form of foreign currency. Tourism continues to be a developing business, especially in the context of globalization and a borderless world, due to economic growth, increased accessibility, and positive relations between countries. (Kastury, 2017).

The use of digital technologies in travel by tourists is a developing trend that the tourism sector must now react to. The growth of the tourism sector today has been affected by these social media components that further promote business transactions between companies offering services to tourists. Furthermore, if the tourism sector wants to maintain its competitiveness in the market, Au (2010) and Jonscher (2011) suggest that it should not disregard the influence of social media. Digitization in aviation, travel, and tourism is predicted to bring billions to the sector's value over the next decade, resulting in higher profits. Customers and the general public benefit from interest of US$700 billion (RM2.8 trillion). This translates to increased tourism revenue, more job possibilities, new enterprises, and infrastructure upgrades, all of which contribute to overall economic growth. Simultaneously, it will contribute to greater cultural understanding and societal unity. Nowadays, cellphones or smartphones are capable of handling the process of reserving hotels and flights, or any other tourism planning. Companies in the industry have changed their marketing strategies to fit the trend of customers doing what they want through their mobile phones by creating "apps" that are suitable for consumers.

As mobile technology in this sector develops, more travellers are using smartphone apps to plan and book their vacations. One of the relevant application characteristics is the capacity to modify every aspect of the travel and tourism industry, including hotel, transportation, and tourist attractions accessible in the user's intended location. For instance, the TripAdvisor app can display reviews left by travellers at the following places. Additionally, users may pick the best hotels for their stay based on their budget, find nearby restaurants and activities, and even compare the price of plane tickets based on their requirements.

The "plateculture" application is primarily created for hardcore food lovers or "foodies" who are looking for a new cultural experience and want to try authentic Malaysian food. The "plateculture" application is simple to use since it provides meal reviews that correspond to the location in the consumer region, as well as price and menu information. The "offpeak" app is for travellers looking for the best deal or restaurant while in the nation. When booking using this "offpeak" programme, users can save up to 50%. Even better, consumers can book without paying in advance. Tourists can use the "CatchThatBus" app to make using public transportation more convenient. This programme will be very beneficial if the user requires a bus ticket to travel to another city from Kuala Lumpur. The "KL Transit" app is much handier because trains are the cheapest mode of transportation in Malaysia when tourists want to travel from one district to another. Users can see when a train is moving and plan accordingly. In Malaysia, the "Grab" or "Uber" application can be used to get to a selected location for a more personal trip.

In order to suit the needs of travellers, the development of mobile applications in the tourism industry will continue to accelerate in the coming years. This is due to the fact that tourists and Malaysians alike prefer to use mobile phones because they make it easier to obtain information or services related to all elements of tourism without wasting time. As the world continues to grow as a result of the global digital revolution, this tendency will affect all tourism businesses.

Conclusion

Governments all around the world are concentrating on new technological applications that save time and effort while also increasing security in their sectors. The use of technology in the tourist and aviation industries is critical to the successful operation of the businesses. Advancements in robotics and applied sciences assist industries in better serving their customers. Simultaneously, the most up-to-date technological equipment allows staff to execute their responsibilities more efficiently and successfully improve the quality of their services. Furthermore, technology enables the tourism and aviation businesses to serve the interests of their passengers and tourists by introducing new technologies that they require. Moreover, technology allows individuals in these areas to focus on telecommuting while simultaneously allowing them to collaborate and share files and crucial information. With the introduction of cutting-edge technological equipment, the industries will find more new passengers and tourists in order to expand their operations and profits. The industry can serve more passengers and tourists in less time than before because of technological advancements. They will have more passengers and tourists than ever before as a result of this.

References

Ali, K. (2008). *Analisis blog sebagai media alternatif penyebaran maklumat dalam industri pelancongan.* Master's Thesis. Johor: Universiti Teknologi Malaysia.

Alrashid, S. A. (2012). Internet adoption in Gulf cooperation council's tourism industry. *New Media and Mass Communication*, 3, pp. 36–42.

Au, A. (2010). *Adoption of Web 2.0 by Tourism Business in NSW.* Retrieved from: http://www.archive.tourism.nsw.gov.au/ (accessed: 10th August, 2021).

AVIATOR (2018). *Air Lanna: A New Airline Startup in Thailand.* Retrieved from: https://newsroom.aviator.aero/air-lanna-a-new-airline-startup-in-thailand/ (accessed: 5th August, 2021).

Bahrom, H., Khalid, M. M. and Yaakob, M. A. Z. (2010). Pelancongan Dari Perspektif Islam: Analisis Pendekatan Fiqh. In Seminar Pengurusan Perhotelan & Pelancongan Islam 2010, Shah Alam, 22nd–23rd June, pp. 52–63.

BH Online (2020). *Malaysia catat RM66.1 bilion pendapatan pelancongan.* Retrieved from: https://www.bharian.com.my/berita/nasional/2020/01/649505/malaysia-catat-rm661-bilion-pendapatan-pelancongan. Februari 2021. Retrieved from: https://www.hmetro.com.my/bisnes/2019/12/524879/airasia-lancar-sistem-pengecaman-wajah (accessed: 10th August, 2021).

Carson, D. (2005). Developing regional tourism using information communications technology. In S. Marshall, W. Taylor and X. Yu (Eds.), *Encyclopedia of Developing Regional Communities with Information and Communication Technology.* Hershey: IGI Global, pp. 176–181.

Chan, N. L. and Guillet, B. D. (2011). Investigation of social media marketing: How does the hotel industry in Hong Kong perform in marketing on social media websites? *Journal of Travel and Tourism Marketing*, 28, pp. 345–368.

Changi Airport Group (2021). *How Robots Are Revolutionalising the Guest Experience at Jewel.* Retrieved from: https://www.changiairport.com/corporate/media-centre/changijourneys/the-changi-experience/how-robots-are-revolutionalising-the-guest-experience-at-Jewel.html (accessed: 10th August, 2021).

Charernnit, K. (2020). Educational and digital tools as factors of service quality increase in Thailand 4.0 era (the case of Suvarnabhumi Airport, Bangkok). *E3S Web of Conferences, 210.* Retrieved from: https://doi.org/10.1051/e3sconf/202021022009 (accessed: 10th August, 2021).

Chuenpukdee, P. (2020). Thammasat business law journal vol.10 2020. *Some Legal Issues of Biometric Data Protection in Thailand*, 10, pp. 119–133.

Constantinides, S. E. and Fountain, S. J. (2007). Web 2.0: Conceptual foundations and marketing issues. *Journal of Direct, Data and Digital Marketing Practice*, 9(3), pp. 231–244.

De Haan, A. R. C. (2010). Sufficient scope in current aircraft technology developments?: A systems analysis application to the multi actor aviation technology system. *European Journal of Transport and Infrastructure Research*, 10(3), pp. 230–248.

Derus, M. A. M. (2019). AirAsia laksana teknologi cam wajah di Melbourne. *B H Online.* Retrieved from: http://www.bharian.com.my/bisnes/korporat/2019/12/636564/air-asia-laksana-teknologi-cam-wajah-di-melbourne (accessed: 10th August, 2021).

Devanesan, J. (2020). Krabi to become first ‘smart airport’ in Thailand with 5G. *Techwire Asia.* Retrieved from: https://techwireasia.com/2020/07/krabi-to-become-first-smart-airport-in-thailand-with-5g/ (accessed: 10th August, 2021).

Ferguson, R. (2008). Word of mouth and viral marketing: Taking the temperature of the hottest trends in marketing. *Journal of Consumer Marketing*, 25(3), pp. 179–182.

Future Travel Experience (2020). *12 Technology Trends for Airlines and Airports to Focus on in 2020.* Retrieved from: http://futuretravelexperience.com/2020/01/12-technology-trends-for-airlines-and-airports-to-focus-on-in-2020// (accessed: 10th August, 2021).

Garrett-Glaser, B. (2019). Thailand’s ambitious plan for aerospace dominance. *Aviation Today.* Retrieved from: https://www.aviationtoday.com/2019/05/28/thailands-ambitious-plans-aerospace-dominance/ (accessed: 10th August, 2021).

Gunelius, S. (2011). *30-Minute Social Media Marketing: Step-By-Step Techniques to Spread the Word About Your Business Fast and Free.* New York: McGraw Hill Companies.

Hawkins, J. M. (2006). *Kamas Dwibahasa Oxford Fajar, Edisi Keempat.* Kuala Lumpur: Oxford Fajar Sdn Bhd, p. 203.

Ibrahim, J. A. and Ahmad, M. Z. (2008). *Perancangan dan pembangunan pelancongan.* Sintok: Universiti Utara Malaysia.

Jawing, O. J. (2020). Usahawan wanita perlu pertingkat kaedah pemasaran berasaskan teknologi. *Utusan Borneo Online*. Retrieved from: https://www.utusanborneo.com.my/2020/03/07/usahawan-wanita-perlu-pertingkat-kaedah-pemasaran-berasaskan-teknologi-abang-johari (accessed: 10th August, 2021).

Jonscher, C. (2011). *Social Media-Potential, Perception and Usage as A Marketing Tool: The Case of Small and Medium Sized Accommodation Providers in Auckland.* Masters Dissertations. Auckland: Auckland University of Technology.

Kansakar, P., Munir, A. and Shabani, N. (2018). Technology in hospitality industry: Prospects and challenges. *IEEE Consumer Electronics Magazine*, 8(3), pp. 60–65.

Kastury, R. (2017). *Kepentingan dan Cabaran Industri Pelancongan kepada Masyarakat negara*. Retrieved from: https://www.kompasiana.com (accessed: 17th August, 2021).

Khairul Nizam, A. M. (2006). *Pembangunan Sistem Maklumat Perjalanan Pelancongan Berasaskan Sistem Maklumat Geografi.* Tesis Sarjana Sains (Geoinformatik). Masters Thesis. Johor Bahru: Universiti Teknologi Malaysia.

Kositchotethana, B. (2014). In-flight WiFi a yawn. *Bangkok Post.* Retrieved from: https://www.bangkokpost.com/thailand/general/439823/in-flight-wifi-a-yawn (accessed: 10th August, 2021).

Litvin, S. W., Goldsmith, R. E. and Pan, B. (2008). Electronic word-of-mouth in hospitality and tourism management. *Tourism Management*, 29, pp. 458–468.

Mokhtar, N. A., Tin, P. B. and Salleh, N. H. M. (2012). Analisis Keperluan Tenaga Manusia dalam Industri Pelancongan di Malaysia. *Prosiding PERKEM*, VII(Jilid 1), pp. 633–640.

Munir, A., Kansakar, P. and Khan, S. U. (2017). IFCIoT: Integrated fog cloud IoT: A novel architectural paradigm for the future internet of things. *IEEE Consumer Electronics Magazine*, 6(3), pp. 74–82.

Noraihan, M. and Ishak, M. S. (2014). Penerimaan media Sosial Dalam Sektor Penginapan Bajet di Malaysia. In International Conference on Culture, Communication and Multimedia Technology (ICON C-COMET'14), Sintok, 29th April, 2014.

Noresah, B. B. (1997). *Kamus Dewan Edisi Ketiga.* Kuala Lumpur: Dewan Bahasa Dan Pustaka, p. 756.

O'Reilly, T. (2005). What is web 2.0? Design patterns and business models for the next generation of software. *Communication and Strategies*, 65(1), pp. 17–37.

Pascu, L. (2019). *Facial Recognition to Replace Passport Checkup at Thailand's Most Crowded Airports.* BIOMETRIC UPDATE.COM. Retrieved from: https://www.biometricupdate.com/201910/facial-recognition-to-replace-passport-checkup-at-thailands-most-crowded-airports (accessed: 10th August, 2021).

Pukdesree, S. and Netinant, P. (2016). Conceptual framework: The adaptive biometrics authentication for accessing cloud computing services using iPhone. *Advances in Intelligent Systems and Computing*, 463, pp. 209–216.

Saifutdinov, F., Jackson, I., Tolujevs, J. and Zmanovska, T. (2020). Digital twin as a decision support tool for airport traffic control. In 2020 61st International Scientific Conference on Information Technology and Management Science of Riga Technical University, ITMS 2020, Riga, 15th–16th October, 2020.

Sinar Harlan (2018). *Penggunaan digital, teknologi majukan industri pelancongan.* Retrieved from: https://www.sinarharian.com.my/article/1860/BERITA/Nasional/Penggunaan-digital-teknologi-majukan-industri-pelancongan (accessed: 14th August, 2021).

Usapein, P. and Chavalparit, O. (2017). A start-up MRV system for an emission trading scheme in Thailand: A case study in the petrochemical industry. *Journal of Cleaner Production*, 142, pp. 3396–3408.

Xiang, Z. and Gretzel, U. (2010). Role of social media in online travel information search. *Tourism Management*, 31, pp. 179–188.

8 A Chaotic or Orderly Digitalization?

Malaysia's Resilient Model for Sustainable Rural Tourism

Farah Hida Sharin, Ilham Sentosa, and Ravindra Kumar Perumal

Introduction

The evolvement of sustainability concepts has become a new global aspiration under the United Nations' Sustainable Development Goals, for which businesses should be committed to implementing mechanisms for sustainability (Naciti, 2019). The United Nations has identified business sectors, along with innovation, as critical components of addressing the challenges of sustainable development (United Nations, 2020a). Due to the widespread effects of technology on society, there has been a widespread shift away from the industrial ages (World Tourism Organization and Global Tourism Economy Research Centre, 2017). To align with the global agenda, the Malaysian government crafted the Shared Prosperity Vision 2030, which appears to serve as a commitment, guiding principle, and enabler for the government to accomplish these goals. This vision aims to make Malaysia a sustainable-growth nation by 2030, while also assuring an equitable distribution for citizens across regions, different categories of income, ethnic groupings, and also the supply chain channels (Ministry of Economic Affairs, Malaysia, 2019).

Since the 1970s, the government has focused on eradicating poverty for all citizens, regardless of ethnicity. As stated in the Tenth Malaysia Plan, poverty incidence has been greatly reduced, from 49.3 percent in 1970 to 3.8 percent in 2009. There are numerous efforts for improving the bottom 40 percent (B40) of households' standard of living, constructing a more progressive society, ensuring access to basic services and infrastructure, increasing Bumiputera economic participation, and promoting corridors as a growth engine (Economic Planning Unit, 2010). Through the generation of income, various support initiatives for micro-enterprises, and the enlargement of human capital, efforts were made to enhance the quality of life of rural households and increase the participation in the economy as reported in the Eleventh Malaysia Plan (Economic Planning Unit, 2015).

The research determined that the national mean Production Linked Incentive Scheme (PLI) for the year 2019 was RM2,208 per month, while the mean PLI

DOI: 10.4324/9781003315179-12

for food was RM1,038 per month and was based on an average family size of 3.9 people. This value is calculated by dividing the total PLI value in each household by the sum of households in Malaysia. The poverty rate in the United States decreased from 7.6 percent in 2016 to 5.6 percent in 2019. Similarly, the poverty rate in Malaysia decreased, in both rural and urban areas, to 3.8 percent and 12.4 percent, respectively. For the present, the rate of extreme poverty in the year 2019 decreased to 0.4 percent from 0.6 percent in the year 2016 (Department of Statistics Malaysia, 2020).

Moreover, with the prior successful poverty-reduction method, the government is constantly trying to eliminate poverty and manage the B40 of households, including the vulnerable and marginalized groups. This is one of the key areas stated in the Shared Prosperity Vision (SPV), which also upholds the responsibility to ensure that, by the year 2030, the Malaysian citizen should be provided with well-being and an adequate standard of living. The definition of an adequate standard of living consists of three elements in a citizen's life, which are, first, being financially able to meet basic needs; second, the capability to participate in community, family, and social activities; and thirdly, the ability to lead a meaningful and dignified life. A decent standard of living will consider the income of a citizen based on expenditures, income classes, and localities that differ from one another. The decision on a reasonable standard of living will guide each individual, worker, business, and government in evaluating and developing programmes to enhance people's socioeconomic conditions (Ministry of Economic Affairs, Malaysia, 2019).

To ensure that the vision outlined above can be realized, one effort has been undertaken to assist rural areas through enhancing tourist sector products targeted primarily at rural tourism. According to the Ministry of Tourism, Arts and Culture, Malaysia (2020), though the National Tourism Policy 2020–2030, the government has crafted a blueprint for tourism transformation based on three key pillars which are sustainability, competitiveness, and inclusiveness. In order to achieve the transformation, six strategies have been established. The Malaysian government should be able to implement the process of transforming the capacity of tourism-related agencies with their tourism core skills as the first strategy. Additionally, a Special Tourism Investment Zone (STIZ) can create a plan to increase the innovation of tourism goods and values in response to the expanding globalisation of the market. The third strategy is to emphasize Smart Tourism by embarking on a digitalization platform to transform Malaysia's tourism industry. The fourth strategy is nurturing demand sophistication by increasing the depth of the tourist experience in support of customized and discerning travel. As the fifth strategy, the National Tourism Policy should be aligning with the agenda of the Sustainable Development Goals of the United Nations by reinforcing the importance of tourism as a catalyst for economic development within inclusive, value of responsible and sustainability. Lastly, the sixth strategy is fostering the service culture via the growth of human capital and competence that naturally cascades throughout the degree of diversity.

Tourism Industry

As one of the largest contributors to economic growth not only in Malaysia but also around the world, the tourism sector supports millions of people's livelihoods and teaches billions more about their own and other cultures, as well as the natural world. In addition, by creating jobs for professionals and experts, notably for immigrants, students, women, and older employees, the tourism industry boosts cities and aids local, coastal, rural, and distant communities (Economic Commission for Latin America and the Caribbean, 2020).

Prior data indicated that the rate of growth for tourist income was 3.8 percent, and that the rate of growth for international visitors in 2011 was 4.6 percent. Additionally, foreign passenger transportation revenues totalled US$196 billion in 2011, increasing total worldwide tourism exports to US$1.2 trillion. As a result, travel and tourism looked to be a big industry in the world, accounting for 9 percent of global GDP. This statistic is greater than the automobile industry's 8.5 percent and somewhat lower than the banking sector's 11 percent, resulting in increased investment and employment creation for 260 million people globally. In 2011, travel and tourism accounted for 4.5 percent of global capital investment, reaching US$650 billion (Scowsill, 2011).

Historically, at the end of the nineteenth century, several major reasons for the increase in tourism and travel included an improvement in humans' living income due to an increase in the process of generating industrial output associated with the era of revolution; affordable prices and easy access to travelling due to transportation technology innovation; annual holidays by the government; changes in views on nature and the environment; and also the desire to travel to get a better education in other countries.

According to Goeldner and Ritchie (2006), the definition of tourism states that tourism as a whole includes all of the processes, activities, and effects resulting from the relationships and exchanges between tourists and tourism suppliers, as well as governments, host communities, and the surrounding environment. Whereas Camilleri (2018) describes the concept of tourism by illuminating the tourism rationale by elucidating the motivation of tourists to have an experience of travelling, and describes the various components that comprise the tourism activity. The destinations visited by tourists are within their reach, and they require lodging if their visit exceeds 24 hours. Both leisure and business travellers have the option of visiting attractions and participating in recreational activities (Osman and Sentosa, 2013b).

Based on the statistics, the growth of tourism in Asia and the Pacific regions was strong in the year 2016. This demonstrates that tourism in the regions of Asia and the Pacific stimulates dynamic economies and should continue to propel the region forward as the main source market for global tourism, a spark that has burned until 2017. International visitors to the Asia-Pacific area accounted for one-fourth of all 2016 international arrivals and for 30 percent of worldwide tourism receipts. In addition, the region welcomed 317 million tourists from outside the region and accounted for about 40 percent of all tourism-related spending worldwide (World Tourism Organization and Global Tourism Economy Research Centre, 2017). The tourism industry has significant potential to climb up the value chain and produce new tourism products and higher-value services with the growth of economic activities in the market.

Foreign tourist activity is expected to provide an increase in revenue of US$ 1.5 billion in 2019 and total of US$ 1.8 billion in the year of 2030 in Asia and the Pacific regions as a whole (World Tourism Organization and Global Tourism Economy Research Centre, 2017). This industry is significantly impacted by the process of technological advancements and innovations. This is evident in its impact on transportation, which is the sector's bedrock. Without technological advancements in larger, speedy, and more fuel-efficient aircraft carriers, the number of passengers travelling would not have increased from 300 million in 1970 to 3.6 billion in 2016.

The other aspect of technology that has an effect on tourism is advancements in the digital realm, particularly in information and communication technology (ICT). The impact of the tech-savvy millennial generation, who are currently in charge of tourism, has led to the fast evolution of the travel sector as a result of ICT, which has changed booking habits. Other than that, the sharing economy has changed the business model of local housing and transportation. Tourism will be profoundly shaped by technological advancements in the future (Al-Jader and Sentosa, 2015).

Regrettably, the tourism field has been significantly harmed by the coronavirus or COVID-19, and the subsequent containment measures. Depending on how severe the crisis was, updated predictions for 2020 indicated that the possible shock led to a 60–80% decline in foreign travel (United Nations World Tourism Organization, 2020a). Due to all the restrictions, the tourism business is one of the sectors most heavily hit by the pandemic of COVID-19, which has wreaked havoc on economies, livelihoods, public services, and opportunities across the globe (United Nations, 2020). By the date of 20 April 2020, 100 percent of tourism destinations around the world had implemented travel limitations. Consequently, this resulted in an estimated 100–120 million job losses and the loss of up to US$1.2 trillion in exports revenue (United Nations World Tourism Organization, 2020b). This pandemic affected this industry significantly more compared to previous crises such as the 2008–2009 financial crisis and also the SARS infection, fundamentally altering the global travel industry. For instance, a number of small Commonwealth states, notably small island developing states (SIDS), rely on foreign tourism for up to 90 percent of exports and a substantial portion of GDP.

According to Tourism Economics' newest estimates, global foreign travel is anticipated to fall by more than 50 percent in 2020, with 2019 levels not being reached until 2023. Tourism's essential nature has transformed, and recovery is expected to be as slow. Even if the virus is confined, it is unlikely that this sector will entirely recover in the foreseeable future. Prospects for recovery are dependent upon the length of the crisis and the amount of time required to recover (Kampel, 2020). Additionally, the viral outbreak's impact on tourism is expected to be asymmetrical and extremely localized within countries, with some locations being disproportionately vulnerable as a result of their reliance on the sector (Organisation for Economic and Co-operation and Development, 2020). Malaysian tourism is likewise not immune to the consequences of COVID-19.

Malaysian Tourism Industry

Malaysia aspires to be among the world's top ten tourism destinations in terms of both arrivals and receipts by 2030 (Ministry of Tourism, Arts and Culture, Malaysia, 2020). Malaysian tourism accounts for 13.3 percent of GDP, amounting to around US$33.6 billion, according to Tourism Satellite Accounting. In Malaysia, the service industries, including tourism, presently account for 46 percent of total revenue and GDP, compared to 30 percent for manufacturing, 9.3 percent for agriculture, and 7.3 percent for mining (Osman and Sentosa, 2013a). Tourism is a rapidly increasing industry that appears to have been well-known and important for a number of years (Musa, 2000). This industry is also a government-supported industry, as seen by the different Malaysian Development Plans (Osman and Sentosa, 2013c). Malaysia is ranked 11th in terms of growth, 35th in terms of absolute size, and 63rd in terms of relative contribution to the country's economy (Sentosa et al., 2021).

Table 8.1 summarizes international tourist arrivals and revenue receipts in Malaysia from 2010 to 2020. Based on this data, the foreign tourist arrivals and receipts for Malaysia are increasing year after year, but due to the global pandemic and travel restrictions imposed by all countries, the number is expected to decline significantly in 2020.

In order to be sustainable, service quality, which serves as a product attraction, is the key to a tourism destination's success (Osman and Sentosa, 2013b).

Table 8.1 The statistics of foreign tourist arrivals and receipts for Malaysia

Year	*Arrivals (million)*	*Receipts (MYR billion)*
2010	24.6	56.5
2011	24.7	58.3
2012	25.3	60.6
2013	25.72	65.4
2014	27.44	72.0
2015	25.72	69.1
2016	26.76	82.1
2017	25.95	82.1
2018	25.83	84.1
2019	26.10	86.1
2020	4.33	12.7

(Source: Ministry of Tourism, Arts and Culture, Malaysia, 2021.)

Sentosa, Sharin, Hadi, and Wahid (2021) reinforce the statement that attractions are the primary motivation for travelling to a destination, with customer satisfaction and customer loyalty relationships serving as complementary factors. In the context of Malaysian tourism, rural products such as batik material, Songket apparel, and also local handicrafts can attract tourists to get to know further and visit Malaysia. There are also resorts and hotels, logistics and transportation, local food and tour guides, scenery and beaches like Perhentian Island, and historical destinations like in Pasir Salak Complex, Perak. Moreover, mega sports or cultural events such as the Formula One Malaysia Grand Prix and Ironman Malaysia could also be considered as tourism products along with rural tourist products that have been popular in recent years. The concept of a homestay in a village setting is popular with both foreign and local tourists. All of these are available to travellers to satisfy their travel goals. The Ministry of Tourism, Arts and Culture, as part of a strategic move, is enhancing the appearance of tourism product categories by adding value to the existing tourist experiences through product development, strengthening enablers and removing impediments to the growth and development of the following products:

- Ecotourism
- Adventure tourism
- Youth tourism
- Sports tourism
- Cultural and heritage tourism
- Community-based tourism
- Island tourism
- Coastal tourism
- Urban tourism
- Shopping tourism
- Medical and wellness tourism
- Meetings, incentives, conferencing, exhibitions (MICE)
- Cross border tourism
- Rural/agri-tourism
- Cruise tourism
- Special interest tourism
- Muslim-friendly tourism

(Source: Ministry of Tourism, Arts and Culture, Malaysia, 2020.)

As stated by Osman and Sentosa (2013b), the growth of tourism in Malaysia is intimately linked to the state's goals of eliminating poverty, rebuilding rural areas, and lessening inequities between the rich and poor, culminating in the coherence of Malaysian culture and, finally, the encouragement of national unity. Some researchers urge that the industry germination should be aided by both the private sectors and local authority body with the target of achieving competitiveness and sustainability (Goeldner et al., 2000). In addition to these factors, trust in the government's capability is a critical component of sustainability. Wong and Lai (2021) investigated the factors that influenced Macao's residents to support the tourism industry during and after the COVID-19 pandemic, and assuredly, it demonstrated that the government's enforcement actions can bolster support especially during the recovery process. Community satisfaction with government action, the study found, increases community trust and support for tourism recovery.

Other than that, the ability to compete effectively and the ability to earn a profit are critical components of tourism sustainability. By contrast, sustainability refers to a destination's capability to protect the quality of environmental resources, physically, socially, and culturally, while competing in the market. Unfortunately, the Malaysian government's analysis indicates that the country's tourism industry has grown complacent. The existing products in Malaysia are becoming stale and unattractive due to the lack of innovation and creativity caused by an excessive reliance on government and a silo mentality. Due to this incapability to cultivate an excellence of working culture, the sector has experienced a real or perceived decline in service quality relative to ASEAN neighbours. Additionally, insufficient destination management strategies have resulted in unsustainably developed physical infrastructure and negative visitor experiences. Finally, Malaysia has lagged behind in adopting Smart Tourism as a result of an over-reliance on traditional marketing and promotion. As a result, the industry must continually reinvent and transform itself to remain competitive (Ministry of Tourism, Arts and Culture, Malaysia, 2020).

Following this, management information systems for the tourism industry are crucial for policy creation, strategic planning, routine decision-making, and overall performance evaluations (Kayat et al., 2008). Four action strategies for Smart Tourism were introduced by the government that involved technology use. These were: first, that the tourism player should optimize e-marketing usage by stepping up synergy between the government and tourism industry players in driving digitalization; second, embracing the knowledge sharing economy to innovate the informal tourism sector and rural tourism; third, leveraging the analytics of big data to support decision-making and future planning; And last, increasing visitor arrivals to rural areas through digitalization (Ministry of Tourism, Arts and Culture, Malaysia, 2020).

Rural Tourism Development in Malaysia

Malaysia introduced rural tourism in 1995. The number of tourism operators in this sector is quite modest, and several of them take benefits from the Malaysian government's different incentives. Rural tourism encompasses a variety of activities in many countries with varying natural environments and cultures in a rural settings, and attracts tourists (Osman and Sentosa, 2013b). Tourists can participate in natural-based activities such as sightseeing, mountaineering, agritourism, cultural tourism, homestays, and others. The various forms of tourism activities conducted in rural areas perplexed industry professionals regarding the proper definition of rural tourism. Previous definitions claimed that this rural tourism only included farm and non-farm tourist activities that took place in rural areas (Oppermann, 1996). However, Oppermann (1997) expanded on this by including activities in mountainous locations. It was previously stated and reaffirmed by the Organisation for Economic Co-operation and Development (OECD) (1994) that rural tourism is a complex varied activity that is not limited to farm-based tourism. It encompasses farm-based vacations, ecotourism and nature, climbing, and horseback riding vacations, adventure, educational travel, arts and heritage tourism, ethnic tourism, and others.

Besides the definition of the term, numerous studies also emphasize rural tourism's significant contribution to the economic and social revival of rural communities (Osman and Sentosa, 2013a), where the prior study includes as well the challenges encountered by farmers, community participation, and the social and psychological effects of tourism (Opperman, 1996). According to Ollenburg and Buckley (2007), while both economic and social reasons drive the expansion of farm tourism in Australia, social motives were shown to be marginally more important than economic motivations in Europe and the United States. The definitional analysis reported four critical aspects of rural tourism which are tourist experiences, the location, sustainable development, and community-based characteristics (Rosalina, 2021). Sustainable rural tourism will lead to improved economic performance based on the review and measurement of factors such human, social, natural, and economic capital which drive successful territorial dynamics in rural areas (Sánchez-Zamora et al., 2014).

Rural tourism's capacity to integrate and enhance the rural economic ecosystem prompted the formation of the Rural Tourist Master Plan (RTMP) in 2001 to plan and steer rural tourism development in the country. Hamzah (2004) stated that this master plan intends to boost rural tourism disbursement from RM350 million to RM1 billion over a five-year period. This will be accomplished by promoting and rebranding rural tourism development in the country in the following ways: by emphasizing the attractive scenery of lush tropical landscapes; by presenting activities and amenities that visitors can participate in and enjoy safely; by offering new and improved accommodation options; and by emphasizing a smiling and friendly customer service approach. According to Hamzah (2004), the RTMP proposed that rural tourist development in Malaysia be transformed in order to achieve significant growth throughout the country. This will involve improvements to rural infrastructure, the building of capacity, and the enlargement of human resources. In addition to the proposed infrastructural improvements, the RTMP aims to raise the class of lodging, upgrade tourist facilities and services, and ensure the ongoing maintenance of products and facilities. As a result, the Malaysian government has budgeted RM100 million to boost rural tourism in the country over a five-year period by quality improvement of tourism products and services offered in rural areas. This endeavour is sustained and supported by the government through the transformation of rural regions in order to improve the rural communities' well-being, as mentioned in the Eleventh Malaysian Plan (Economic Planning Unit, 2015). 26.3 percent of Malaysians, or 7.92 million people, reside in rural districts, which vary in their proximity to peri-urban areas, social amenities, infrastructure, and economic prospects. Enhancing rural communities' well-being remains a priority in the Eleventh Plan, consistent with the goal of achieving balanced growth across geographies. Five strategies were implemented throughout the Eleventh Plan to accomplish these goals: extending the basic infrastructure of rural provision by streamlining planning and increasing road, water, and electricity supply; encouraging investment by creating a more conducive business environment; improving rural-urban linkages by increasing connectivity and mobility, as well as to strengthen the role of leveraging ICT; expanding the access to basic services by expanding the mobile communications network (Economic Planning Unit, 2015). All of these measures contribute to the growth and sustainability of rural tourism. Numerous studies have classified rural tourism activities into numerous categories, but this research concentrates on four unique types of rural tourism in Malaysia: homestay tourism, camp or caravan site tourism, adventure tourism, and ecotourism (Othman et al., 2013).

Homestay Tourism

The homestay programme was established by the Ministry of Culture, Arts, and Tourism in 1988 as an alternative for tourists looking for lodging in rural areas (Kayat et al., 2008). Tourists who engage in the programme get the opportunity to experience daily life with host families, allowing them to learn about the hosts' culture and traditions. This programme style has been confirmed by Mohamad Nor (2019), who stated that tourists are drawn to homestays for the opportunity to immerse themselves in culture and heritage. It can be inferred that homestays are synonymous with cultural tourism, centred on the pursuit of and involvement in novel and profound cultural experiences, whether psychological, aesthetic, intellectual, or emotional. A majority of cultural activities are conducted by rural tourist operators (Harris, 2009), who display indigenous cultural communities' traditions through festivals, rituals, values, and ways of life. The majority of homestays are run by villagers with assistance from the state government and tourism organization. The Desa Murni Homestay Program was one of the initial programmes, involving five communities near Temerloh, Pahang. The cohesion among the residents of these communities who had worked painstakingly to build a successful homestay programme had a tremendous influence on the local economy. Another successful homestay programme in Malaysia is the Kampung Relau and Sungai Haji Dorani Homestay in Selangor. Occasionally, however, homestay lodgings created particularly for the purpose of attracting tourists to remote locations are unable to provide enough amenities and a comfortable setting with sufficient privacy (Hawariyuni et al., 2018). The majority of homestays allow guests to stay with the owner's family members, with no physical barriers between them. Additionally, Mohamad Nor (2019) identified many issues and problems such as unethical, unregistered homestay operators that may stain the good name of registered homestays as well as a lack of homestay regulations and policies monitored by local government. Other than that, as Richards (2018) points out, cultural tourists are no longer following the attractions; rather, they are constructing their own definitions of what constitutes the usage of culture. They seek out places that are distinctive, which distinguishes them as tourists and thus as individuals.

Camp or Caravan Site Tourism

Camping is characterized by sleeping in a tent in the outdoors as a low-cost vacation choice (Blichfeldt and Mikkelsen, 2013). Camping has been mentioned in numerous ways in tourist research pertaining to second homes, scenic-drive tourism, nature, and so on (Blichfeldt and Mikkelsen, 2014), and many campers have recently shifted to a different sort of holiday, one that includes one-night stays in any recreational vehicle (RV), cabin, caravan, or another type of temporary shelter (Brooker and Joppe, 2014). Theoretically, one may conclude that this tourism activity encompasses both lodging and outdoor recreation equally (Rogerson and Rogerson, 2020). Most camp site operations are located in rural areas or nature environments and so are able to develop the economy of rural communities (Sentosa et al., 2021). This is emphasized as well by Brooker and Joppe (2014), who claimed that camping has grown beyond its origins as a low-cost, transitory stopover, growing into a highly fragmented niche tourism business. However, the growing demand for luxurious outdoor experiences, dubbed glamorous camping or "glamping," is a new niche form of tourism that Tourism Malaysia is eager to promote (Tourism Malaysia, 2015). This new niche form of tourism validates the fact that camping tourism is evolving away from traditional low-cost modes of accommodation and towards more expensive modes of accommodation (Rogerson and Rogerson, 2020). While the majority of Malaysian glamping options, such as Caravan Serai and Tanah Aina in Pahang, emphasize comfort and convenience, others, such as Sekeping Serendah in Rawang, Selangor, with its mud and glass houses set within a small forest with nearby waterfalls, and Japamala Resort on Tioman Island, with their romantic treetop "love-nests" overlooking the blue seas below, offer more deluxe experiences. Nevertheless, Mikulić, Prebežac, Šerić, and Krešić (2017) found that camping and caravan parks are "a somewhat overlooked segment of the tourism and hospitality industries." There are numerous emerging literatures in South Africa, which should serve as a signal for several concerns that should be explored as part of a future study agenda (Rogerson and Rogerson, 2020).

Adventure Tourism

Adventure tourism is becoming one of the most popular outdoor activities in the entire world. This is a global phenomenon as individuals become more concerned about their health and desire to live a healthy lifestyle (Isa et al., 2015). Adventure activities are classified into "soft" and "hard" adventures. Soft adventures involve less risk, provide more comfort in accommodations, and are less physically demanding. Soft adventure tourists may also use mechanized transport such cars and motorbikes. Varley and Semple (2015) elaborated on this point, stating that soft or gentle adventure is a concept centred on "feeling, sensing, and investing in location, community, belonging, sociality, as well as time-honored and natural traditions." Horseback riding, snorkelling, canoeing or kayaking, and wandering in locations of remarkable natural beauty are all examples. On the other hand, hard adventures, such as backpacking or river journeys, usually require extremely basic equipment, a higher risk factor, and increased physical exertion. 4×4 exploration, bungee jumping, caving, mountain climbing, paragliding, and rock climbing are all examples of extreme adventure tourism that are gaining popularity in the market today. There are various adventure tourism activities in Malaysia, including paragliding in Kuala Kubu Baru, scuba diving in Borneo Island, tandem skydiving in Kuala Lumpur, parasailing in Langkawi Island, white water rafting in Kampar, bungee jumping in Sunway Lagoon Theme Park, and caving in Gua Tempurung. The high demand for adventure tourism has drawn academics, practitioners, and policymakers to organize colloquia, conferences, talks, seminars, and workshops on this subject. The majority of dictionaries define adventure tourism as "a unique experience that involves some element of danger and uncertainty." Numerous tourism researchers agree that the tourist industry's very nature is geared towards creating positive and memorable experiences (Pine and Gilmore, 1998; Nielsen, 2004; Isa et al., 2012), as tourism operators such as commercial lodging, airlines, and travel agents charge customers for tangible services, goods, or commodities. Additionally, other studies emphasize how adventure tourism will become a component of rural tourism. From a destination management standpoint, adventure tourism, such as hiking, is a unique product since it improves rural development by providing year-round tourism and so helps offset off-peak seasons (Kastenholz and Rodriguez, 2007; Wall-Reinius and Back, 2011). Additionally, being an adventure tourist has a lower detrimental impact on the environment than other intrusive activities (Thongdejsri and Nitivattananon, 2019). This is reinforced by allowing multiple day trips and presenting a variety of options with associated services along the route, all of which contribute to sustainable tourism (Olmos-Raya et al., 2018).

Ecotourism

Ecotourism is currently trendy as travelling activity that involves the attractions of nature, places a premium on nature preservation, and requires tourism activities to have a minimal impact on the degradation of the environment while also respecting indigenous cultures and ways of life (Wallace, 2019). Ecotourism as a niche in tourism has indeed flourished and made a name for itself in academia and the industry over the last two decades. The very principle of ecotourism, which seeks to strike a sustainable balance between protecting and conserving natural resources and promoting community well-being, has unquestionably garnered enormous positive support on a global scale. This has resulted in a plethora of conferences, grants, projects, and discussions within the global ecotourism community, with numerous exemplary success stories despite a variety of obstacles.

There are several established ecotourism destinations in Malaysia, including the Forest Research Institute Malaysia's FRIM Canopy Walkway, which is one of the main attractions for visitors (FRIM). Since its inception in 1992, the walkway has attracted visitors from around the world, generating significant revenue for FRIM's management. Additionally, a UNESCO World Heritage Site in Sabah, Borneo, called Kinabalu Park has become one of the hotspots of the region's biodiversity (Sheena et al., 2015). Besides this, there are several proposals and ongoing projects of ecotourism in Malaysia. The Perak River Tourism Corridor (PRTC) is an area along the river that has been proposed as the state's future tourism focal point, and Mossy Forest Park in Cameron Highlands, Pahang, is another example (Aziz et al., 2015). Sheena, Manohar, and Azlizam (2015) discovered that it is critical for ecotourism operators to segment the ecotourists into several groups. They can be grouped by their interest in the primary ecotourism attributes and also previous and current experience. Three categories of ecotourist can be identified which are soft, structured, and hard ecotourists by the analysing and segmenting process. Ecotourism operators are able to meet ecotourists' expectations and analyse the trip characteristics of ecotourists for services and preferred travel arrangements. This was supported previously by Weaver and Lawton (2002), who significantly aided in clarifying and enhancing the understanding of the segmentation of the ecotourist.

Conceptual Development

To create a conceptual model for this study, two key theories, namely the resource-based view (RBV) and the theory of diffusion of innovation (DOI), were applied in this research. Community resources tourism (CRT) is an exogenous variable (IV) which includes physical resources, cultural resources, human resources, and financial resources, and which can be associated with RBV since Penrose (1959) stated that having, deploying, and effectively using resources will provide more results than any other industrial organization. By this approach, the firm's sustainable competitive advantage in terms of its resource performance was defined, and he invented the term "resource base view" to describe how he viewed firms from the perspective of vital resources. Researchers emphasize on the definition of "core competency," and it placed a premium on a variety of important resources (Prahalad and Hamel, 1990), dubbed the company capability, which they referred to as competitive advantage (Barney, 1991).

Even today, the RBV still seems to be a relevant theory that can be practised and applied since it is an extension of the firm-growth theory (Barney, 1991; Grant, 1991). From a management perspective, the sustainable development dimensions should be centred together with the enhancement and comprehension of the study's setting and approaches. It is critical to consider a resource-based perspective. The evaluation of all resources enables businesses to operate more efficiently. By choosing a suitable growth plan, tourism business owners can overcome growth adversities or minimize direction changes and growth challenges. This theory states that the managerial team needs a structure to explain how a firm's innovation and openness characters and evolve into quick and timely, or influence towards the dynamic of the marketplace and flexibility as well, and this was mentioned by Kiiru (2015) and Teece, Pisano, and Shuen (1997). In this context, CRT dimensions are predicted to be a primary influencer in the sustainability of rural tourism, as well as successful business performance ultimately.

Along with that, all dimensions of resources at the same time are related to the community's perception towards tourism (Andereck et al., 2005). The way tourism operators strategize to use resources in the community will create a discrete competency and positive impact on the tourism industry and local economic development (Sentosa et al., 2021). The RBV also found chances for firms to become more competitive in any economic conditions by emphasizing the uniqueness of the resource (Grewal et al., 2011). Furthermore, RBV theory also concentrates on central concerns relating to capacities or company performance and relevance in the field of strategic management (Darcy et al., 2014). To ensure rural tourism's sustainability, the achievement of new strategic goals should be developed by the act of merging available resources (Gundry et al., 2011), which aligns with the dimension of CRT (*exogenous variables*) in this present research. Combining all resources may enable firms to strategize the use of resources in order to enhance the service performance, particularly for tourism businesses working on a shoestring budget.

Correspondingly, the mediator variable for the present research is the technology development process (TDP) which is comprised of the dimensions of participation, entrepreneurship, and empowerment. This transformation variable is influenced by the DOI theory introduced by Rogers (1995) and frequently uses the terms "technology" or "innovation" as synonyms for "diffusion of innovation." Diffusion can be defined as "a process through which innovation is thoroughly disseminated among members of the social system over time via established routes" (Rogers, 2003). His idea has been widely included in the theoretical framework for technological acceptance and has served as a foundation for numerous studies on a wide variety of issues (Al Mamun, 2018). This DOI theory underpins the construction of this study model and its applicability to the current state of Malaysian tourism where numerous analyses suggest a dearth of technology adoption in the formulation of business strategies. According to Rogers (2003), technology and innovation are essential for the sustainability of rural tourism in Malaysia, and innovation is defined as a notion, practise, or effort that is recognised as particular to one entity or a group of other adoption components. The components of these variables such as participation, entrepreneurship, and empowerment will complete the technology development process. Johnson (2015) identifies a variety of domains such as the management field, communications channels, political science, healthcare management, entrepreneurial innovation, and many more in which this theory of diffusion of innovation can be applied. The four fundamental factors of innovation spread are invention, communication channels, time, and social structure (Chege and Wang, 2020). These factors aided the TDP dimension by promoting the invention of entrepreneurial elements among tourism operators, communication channels for community participation, and a social structure that enabled the process of technological development.

Moreover, the negative repercussions of tourism can undoubtedly be mitigated among community members and become a catalyst by transforming tourism. Innovation is fostered by organizational culture (Do et al., 2018). Indeed, innovation is associated with behaviours that demonstrate an affinity for respect and trust, good communication, risk-taking characteristics, freedom of expression, teamwork focus, and some incentive for advancement, and a workplace environment involving team meetings, good relations among employees, and employee empowerment will influence their effectiveness, and up-to-date model improvements (Lijauco et al., 2020; Rogers, 2003, 1995). All in all, innovation places a premium on the evolution of the current concept and on the contribution to meeting economic needs aligned with the United Nations Sustainable Development Goals Agenda 2030 (United Nations, 2015; Francke and Alexander, 2019). The dissemination of innovation elements requires building innovation cultures capable of boosting the effectiveness of innovative competitive advantages that enable tourism enterprises to thrive sustainably in a new market dynamic. The community would embrace and support tourism activities that benefit them, whether economically, socially, or environmentally. Thus, the current research attempts to define CRT in terms of its contribution to rural tourism sustainable performance (RTS) and the contingent role of TDP through the lens of RBV and DOI.

Integrated Empirical Model

The present study initiates the development of an integrated rural tourism sustainability model on how the community development process takes place in rural tourism activities as illustrated in Figure 8.1. This integrated empirical model is based on the exogenous-intervening-endogenous model which was introduced by Deacon and Firebaugh (1988). This model is based on systems theory which is a holistic approach used to understand and explain complex and organized phenomena where each component has a strong relationship and contributes to the whole system (Berger and Jurkovic, 1984). According to Deacon and Firebaugh (1988), the three components in the model which are exogenous, intervening, and endogenous are suitable and applicable to all types of management systems. Each component has its own uniqueness in terms of managing the use of resources and meeting its distinct objectives. This model has been frequently utilized in resource management and finance, according to Goldsmith (2005), while Pennings, Keman, and Kleinnijenhuis (2006) used it to illustrate the political system approach.

This integrated rural tourism sustainability model (Figure 8.1) has been formed to understand the relationship in the community that uses community development as a mechanism for the sustainability of rural tourism. The three

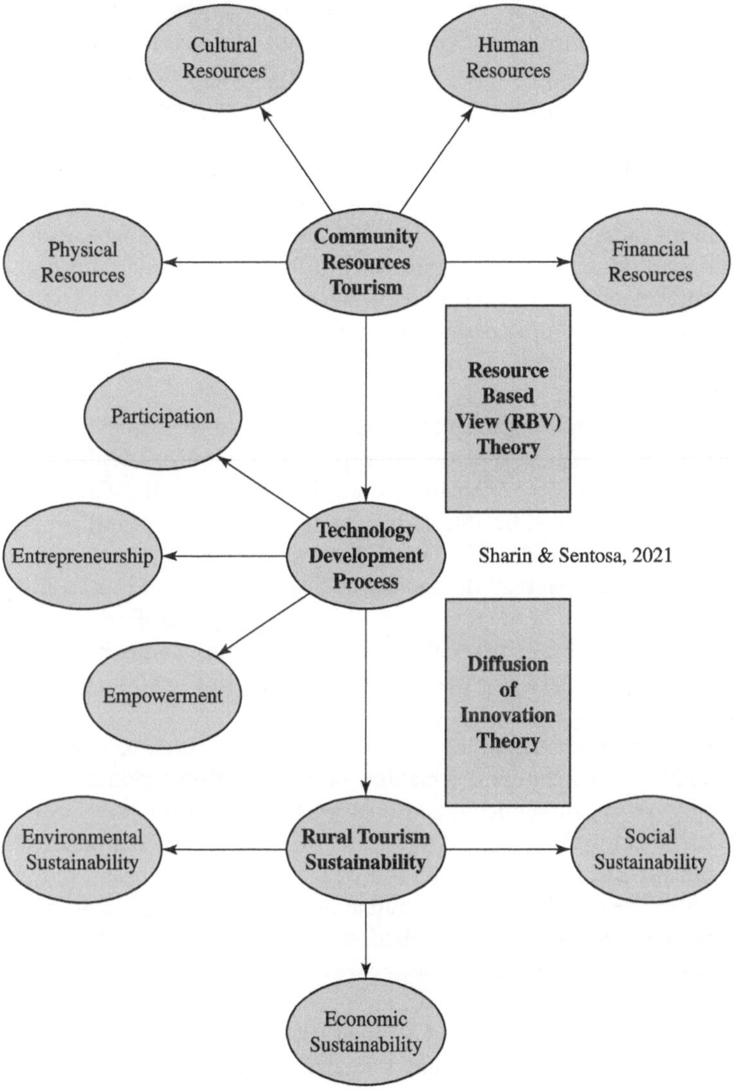

Figure 8.1 Conceptual development of rural tourism sustainability (source: developed by the authors, 2021).

components in this model, namely exogenous, intervening, and endogenous, explain how community development influences and contributes to rural tourism's sustainability. To fulfil the purpose of this present research, the exogenous-intervening-endogenous model has been adapted based on the combination of elements and supported by numerous previous studies. The first component (*exogenous*) comprises the CRT effect on the physical and social aspects of the rural areas. It consists of the four main dimensions which are financial resources, human resources, cultural resources, and physical resources which are all considered as community input. The second component (***intervening***) comprises of community transformation by TDP with an emphasis on the dimensions of participation, entrepreneurship, and empowerment. While the third component (***endogenous***) is called RTS that measures three dimensions: the economic sustainability, the social sustainability, and the environment sustainability. This variable was adapted and re-designated after Andereck, Valentine, Knopf, and Vogt (2005) identified three key areas of tourism's influence on the community: economic, socio-cultural, and environmental. The proper implementation of rural tourism would certainly lead to effective coordination, increased tourist visitation, greater participation, heightened family unity, and supplementary income which are all essential for the success of rural tourism.

According to Chai-Arayalert (2020), leaning promotes ecotourism in destinations and enables the ecotourist to know and experience ecotourism destinations by introducing the application of smart-learning in modern media. Thus, this smart-learning application will be able to remove the barriers for young people to understand ecotourism products by tailoring the specific needs to their online learning platform. Aside from that, the tourism operators in Malaysia lack creativity and entrepreneurial skills to adapt to tourism disruption, specifically digitalization facilities such as internet access, online booking platforms, and e-marketing (Mohamad Nor, 2019). This factor was previously pointed out by Croes (2006), where the homogeneity of tourism products without creativity will be a big challenge in industry competitiveness. Following Kastenholz, Carneiro, Marques, and Lima (2012), visitors choose destinations or rural tourism products based on the breadth, variety, authenticity, and quality of the "tourism experience." Moreover, Thahir, Hadi, Zahra, Arif, Murad, and Mashur (2021) identified that the tourism industry's workforce is not yet sufficiently trained, which has an effect on service quality. The certification of tourism industry employees can contribute to increased visitor satisfaction. The standardization of skills training should be enhanced to ensure that the workforce's awareness and motivation for sustainable tourism are maintained (Sentosa et al., 2021).

Cooper, Fletcher, Fyall, Gilbert, and Wanhill (2005) argued that small and medium-sized firms (SMEs) are becoming a weakness at the micro level due to their lack of capability to manage the supply and demand within the current situation of economic scale. Certain transnational corporations have recognized this and are developing networks. However, this relationship is unable to pursue on a long-term basis and appears to have waned in importance. Given the critical nature of rural tourist development to rural community development, it is critical to perform comprehensive study. The role of government is important in supporting and providing information on the current economic conditions (Rieznik and Lee, 2018). It is necessary to foster organic agriculture development and provide local governments with greater authority in order to achieve sustainable agriculture and rural area development.

Recommendations and Conclusion

The researchers at the outset attempted to investigate Malaysia's policy on rural tourism, configured the strengths, weaknesses, opportunities, and threats in existing rural tourism destinations in Malaysia, explored the best configuration to enhance and promote this product locally and internationally by attracting tourist segments, and recommended lasting policies to government agencies on rural tourism sustainability which could make a viable substitute for declining rural economies in enhancing the gross national product (GNP) of Malaysia. The proposed integrated model (Figure 8.1) will be beneficial to the Ministry of Tourism, Arts and Culture, Malaysia, as well as the Ministry for Rural Development in their strategies on investment and marketing for rural development. Malaysia has immense potential to enhance its presence in the global tourism market, and any improvement recommended if implemented will make significant strides in developing the tourism industry and educating the locals on these issues.

On top of that, the crisis situation created by the onset of COVID-19 exacerbates the difficulty of sustaining rural tourism. The tourism industry is the most severely impacted by this crisis. The proposed integrated model serves as the drivers of economic performance are largely determined by a complex interplay of internal and external forces of rural tourism through digitalization and the impact of technology usage. Additionally, this study will certainly provide a deeper understanding of the relationships between the major factors affecting sustainable rural tourism development for future research. Notably, this model could be used as a comprehensive approach for measuring rural economic performance and addressing rural issues and challenges, as well as being critical for designing technology policy specifically for rural tourism activities in Malaysia.

By doing this, it is hoped to trickle down to the local communities and consequently improve their economic standing especially in the course of recovery from the COVID-19 pandemic. But the question then becomes: are the rural communities ready to take on the challenges posed by tourism activities, and are they prepared to change some of their ways of doing things? In other words, to what extent are they capable and ready to get involved? As with many top-down development strategies and programmes, local communities are expected to carry out the implementation while their capability and readiness are still not fully realized. Do policymakers, tourist businesses, government officials, local authorities, and community drivers play a significant role in ensuring that all the efforts and activities that provide to the community are aligned with the community's objectives, or are we at a chaotic or orderly stage? Is there anyone who can affirm that this endeavour is truly improving well-being, taking into account community carrying capacity, and maintaining its long-term viability for present and future generations?

Future academics interested in rural tourist products and their viability will receive significant insight from this paper's findings, which will be extremely useful for similar study efforts. A system dynamic approach is strongly recommended for further rural tourism ecosystem analysis using latent and observed variables as configured in Figure 8.1. Academicians could use their presentations to publicize the findings of this investigation. Policymakers can utilize the facts in this study to justify the formulation of future laws and regulations controlling rural tourist activities. Rural communities interested in exploring rural tourism as a potential source of revenue and community growth may use the model provided at the conclusion of this research as a benchmark for best practices.

References

Al-Jader, R. A. and Sentosa, I. (2015). A conceptual development on the mediating role of E-service recovery on the relationship between customer determinants and customer retentions in the airline industry in Malaysia (a structural equation modelling approach). *Indian Journal of Commerce and Management Studies*, VI(1), pp. 98–103.

Al Mamun, A. (2018). Diffusion of innovation among Malaysian manufacturing SMEs. *European Journal of Innovation Management*, 21(1), pp. 113–141.

Andereck, K. L., Valentine, K. M., Knopf, R. C. and Vogt, C. A. (2005). Resident perceptions of community tourism impacts. *Annals of Tourism Research*, 32(4), pp. 1056–1076.

Aziz, A., Ajuhari, Z. and Bidin, S. (2015). Recreation and tourism resource assessments along the Perak river tourism corridor. In M. Mariapan, Lin, E. L. A., Bidin, S., Zawawi, A., Abidin Z. A. Z. and Jumaat, N. J. (Eds.), *Ecotourism in Malaysia: Current Scenario*. Selangor: Universiti Putra, pp. 1–13.

Barney, J. B. (1991). Firm resources and sustained competitive advantage. *Journal of Management*, 17(1), pp. 99–120.

Berger, M. and Jurkovic, G. J. (1984). *Practicing Family Therapy in Diverse Settings*. San Francisco: Jossey-Bass.

Blichfeldt, B. S. and Mikkelsen, M. V. (2013). Vacability and sociability as touristic attraction. *Tourist Studies*, 13, pp. 235–250.

Blichfeldt, B. S. and Mikkelsen, M. V. (2014). Camping, tourism. In J. Jafari and H. Xiao (Eds.), *Encyclopedia of Tourism*. Cham: Springer, https://doi.org/10.1007/978-3-319-01669-6_468-1

Brooker, E. and Joppe, M. (2014). A critical review of camping research and direction for future studies. *Journal of Vacation Marketing*, 20(4), pp. 335–351.

Camilleri, M. A. (2018). *Travel Marketing, Tourism Economics and the Airline Product*. Cham: Springer Nature.

Chai-Arayalert, S. (2020). Smart application of learning ecotourism for young eco-tourists. *Cogent Social Sciences*, 6(1), p. 1772558.

Chege, S. M. and Wang, D. (2020). The influence of technology innovation on SME performance through environmental sustainability practices in Kenya. *Technology in Society*, 60, p. 101210.

Cooper, C., Fletcher, J., Fyall, A., Gilbert, D. and Wanhill, S. (2005). *Tourism Principles and Practices*. Harlow: Pearson Education.

Croes, R. R. (2006). A paradigm shift to a new strategy for small island economies: Embracing demand side economics for value enhancement and long-term economic stability. *Tourism Management*, 27, pp. 453–465.

Darcy, C., Hill, J., McCabe, T. J. and McGovern, P. (2014). A consideration of organisational sustainability in the SME context: A resource-based view and composite model. *European Journal of Training and Development*, 38(5), pp. 398–414.

Deacon, R. E. and Firebaugh, F. M. (1988). *Family Resource Management: Principles and Applications.* Boston: Allyn and Bacon.

Department of Statistics Malaysia (DOSM) (2020). *Household Income and Basic Amenities Survey Report 2019.* Retrieved from: https://www.dosm.gov.my/v1/index.php?r=column/ctwoByCat&parent_id=119&menu_id=amVoWU54UTl0a21NWmdhMjFMMWcyZz09 (accessed: the 1st November, 2021).

Do, H., Mazzarol, T., Soutar, G. N., Volery, T. and Reboud, S. (2018). Organisational factors, anticipated rents and commercialisation in SMEs. *International Journal of Innovation Management*, 22(2), p. 1850018.

Economic Commission for Latin America and the Caribbean (ECLAC) (2020). *Building a New Future: Transformative Recovery with Equality and Sustainability (LC/SES.38/3-P/Rev.1), Santiago.* Retrieved from: https://www.pmo.gov.my/dokumenattached/RMK/RMK10_E.pdf (accessed: the 16th November, 2021).

Economic Planning Unit (EPU) (2010). *Malaysia Prime Minister's Department Tenth Malaysia Plan 2011–2015, Putrajaya 2010.* Retrieved from: https://www.pmo.gov.my/dokumenattached/RMK/RMK10_E.pdf (accessed: the 17th November, 2021).

Economic Planning Unit (EPU) (2015). *Prime Minister's Department, Malaysia, Eleventh Malaysia Plan 2016–2020 Anchoring Growth on People.* Retrieved from: https://www.epu.gov.my/ms (accessed: the 17th November, 2021).

Francke, E. and Alexander, B. (2019). Entrepreneurial development in South Africa through innovation: A model for poverty alleviation. *Acta Commercii*, 19(1), pp. 1–11.

Goeldner, C. R. and Ritchie, J. R. B. (2006). *Tourism: Principles, Practices, Philosophies,* Tenth Edition. Hoboken: John Wiley and Sons, Inc.

Goeldner, C. R., Ritchie, J. R. B. and MacIntosh, R. W. (2000). *Tourism: Principles, Practices, Philosophies.* New York: John Wiley and Sons, Inc., pp. 615–665.

Goldsmith, A. (2005). Police reform and the problem of trust. *Theoretical Criminology*, 9(4), pp. 443–470.

Grant, R. M. (1991). The resource-based theory of competitive advantage: Implications for strategy formulation. *California Management Review*, 33(3), pp. 114–135.

Grewal, D., Iyer, G. R., Javalgi, R. R. G. and Radulovich, L. (2011). Franchise partnership and international expansion: A conceptual framework and research propositions. *Entrepreneurship Theory and Practice*, 35(3), pp. 533–557.

Gundry, L. K., Kickul, J. R., Griffiths, M. D. and Bacq, S. C. (2011). Creating social change out of nothing: The role of entrepreneurial bricolage in social entrepreneurs' catalyst innovations. *Social and Sustainable Entrepreneurship*, 13, pp. 1–24.

Hamzah, A. (2004). Policy and planning of the tourism industry in Malaysia. In Proceeding of The 6th ADRF General Meeting 2004, Bangkok, 7th–8th June, 2004.

Harris, R. (2009). *ACT Made Simple: An Easy-To-Read Primer on Acceptance and Commitment Therapy.* Oakland: New Harbinger.

Hawariyuni, W., Rashid, A. Z. A., Alaeddin, O., Krishnan, K. S., Sentosa, I. and Nugraha, Y. M. (2018). Establishing Shari'ah compliant Homestay in Indonesia: Issues and challenges. *American Scientific Publishers Advanced Science Letters*, 24(1), pp. 289–292.

Isa, S. S., Ali, A. Yusof, W. Z. M. and Isa, S. S. (2012). The adaptation of creativity in museum sector: A case study of Malacca Maritime Museum, Malaysia. In 2012 IEEE Symposium on Business, Engineering and Industrial Applications, Bandung, 23rd–26th September, 2012.

Isa, S. S., Hasbullah, R. and Nasir, M. N. M. (2015). Adventure and ecotourism in Malaysia. In M. Mariapan, S. S. Isa, N. A. A. Aziz, E. L. A. Lin and K. R. Hakeem (Eds.), *Adventure and Ecotourism in Malaysia*. Selangor: Universiti Putra, pp. 1–25.

Johnson, M. P. (2015). Sustainability management and small and medium-sized enterprises: Managers' awareness and implementation of innovative tools. *Corporate Social Responsibility and Environmental Management*, 22(5), pp. 271–285.

Kampel, K. (2020). COVID-19 and tourism: Charting a sustainable, resilient recovery for small states, trade hot topics. *The Commonwealth*, 163, pp. 1–14.

Kastenholz, E. and Rodrigues, Á. (2007). Discussing the potential benefits of hiking tourism in Portugal. *Anatolia*, 18(1), pp. 5–21.

Kastenholz, E., Carneiro, M., Marques, C. and Lima, J. (2012). Understanding and managing the rural tourism experience - The case of a historical village in Portugal. *Tourism Management Perspectives*, 4, pp. 207–214.

Kayat, K., Ashikin, N. and Amin, M. (2008). *Penglibatan Komuniti dalam pelancongan lestari, dalam Yahaya Ibrahim, Sulong Mohamad and Habibah Ahmad: Pelancongan Malaysia: isu pembangunan, budaya, komuniti dan persetempatan*. Kedah: Penerbit UUM, pp. 159–170.

Kiiru, G. W. (2015). *Dynamic Capabilities, Strategic Orientation and Competitive Advantage of Small and Medium-retail Enterprises in Kenya*. Doctor of Philosophy Thesis. Nairobi: Jomo Kenyatta University Kenya.

Lijauco, F., Gajendran, T., Brewer, G. and Rasoolimanesh, S. M. (2020). Impacts of culture on innovation propensity in small to medium enterprises in construction. *Journal of Construction Engineering and Management*, 146(3), p. 04019116.

Mikulić, J., Prebežac, D., Šerić, M. and Krešić, D. (2017). Campsite choice and the camping tourism experience: Investigating decisive campsite attributes using relevance-determinance analysis. *Tourism Management*, 59, 226–233.

Ministry of Economic Affairs, Malaysia (2019). *Shared Prosperity Vision 2030*. Retrieved from: https://www.epu.gov.my/sites/default/files/2020-02/Shared%20Prosperity%20Vision%202030.pdf (accessed: the 5th August, 2021).

Ministry of Tourism, Arts and Culture, Malaysia (2020). *National Tourism Policy 2020 –2030 Executive Summary*. Retrieved from: https://www.tourism.gov.my/files/uploads/Executive_Summary.pdf (accessed: the 10th August, 2021).

Ministry of Tourism, Arts and Culture, Malaysia (2021). *Malaysia Tourism Statistics in Brief*. Retrieved from: https://www.tourism.gov.my/statistics (accessed: the 26th August, 2021).

Mohamad Nor, S. B. (2019). *Challenges Faced by Registered Homestay Operators from the Perspectives of Selected Stakeholders in Selangor, Malaysia*. Doctoral Thesis. Selangor: Universiti Putra Malaysia.

Musa, G. (2000). Tourism in Malaysia. In M. C. Hall and S. Page (Eds.), *Tourism in South and South East Asia: Issues and Cases*. Auckland: Butterworth- Heinemann, pp. 144–156.

Naciti, V. (2019). Corporate governance and board of directors: The effect of a board composition on firm sustainability performance. *Journal of Cleaner Production*, 237, p. 117727.

Nielsen, T. (2004). *Understanding the Experience Industry: A Swedish Perspective on Creativity*. Gothenburg: Majornas Copy Print.

Ollenburg, C. and Buckley, R. (2007). Stated economic and social motivations of farm tourism operators. *Journal of Travel Research*, 45(4), pp. 444–452.

Olmos-Raya, E., Ferreira-Cavalcanti, J., Contero, M., Castellanos, M. C., Giglioli, I. A. C. and Alcañiz, M. (2018). Mobile virtual reality as an educational platform: A pilot study on the impact of immersion and positive emotion induction in the learning process. *Eurasia Journal of Mathematics, Science and Technology Education*, 14(6), pp. 2045–2057.

Oppermann, M. (1996). Visitation of tourism attractions and tourist expenditure patterns - repeat versus first-time visitors. *Asia Pacific Journal of Tourism Research*, 1(1), pp. 61–68.

Oppermann, M. (1997). First-time and repeat visitors to New Zealand. *Tourism Management*, 18(3), pp. 177–181.

Organisation for Economic Cooperation and Development (OECD) (1994). *General Distribution Tourism Strategies and Rural Development, Paris 1994 OCDE/GD (94)49, 013927*. Paris: OECD.

Organisation for Economic and Cooperation and Development (OECD) (2020). *Tackling Coronavirus: Tourism Policy Responses*. Retrieved from: https://www.oecd.org/coronavirus/policy-responses/tourism-policy-responses-to-the-coronavirus-covid-19-6466aa20/ (accessed: the 10th August, 2021).

Osman, Z. and Sentosa, I. (2013a). Influence of customer satisfaction on service quality and trust relationship in Malaysian rural tourism. *Business Management Quarterly Review*, 4(2), pp. 12–25.

Osman, Z. and Sentosa, I. (2013b). Mediating effect of customer satisfaction on service quality and customer loyalty relationship in Malaysian rural tourism. *International Journal of Economics Business and Management Studies*, 2(1), pp. 25–37.

Osman, Z. and Sentosa, I. (2013c). A study of mediating effect of trust on customer satisfaction and customer loyalty relationship in Malaysian rural tourism. *European Journal of Tourism Research*, 6(2), pp. 191–206.

Othman, F., Sazali, F. and Mohamed, B. (2013). Rural and community based tourism development in Malaysia: Prospects for homestays as a social economy enterprise. *TEAM Journal of Hospitality and Tourism*, 10(1), pp. 65–76.

Pennings, P., Keman, H. and Kleinnijenhuis, J. (2006). *Doing Research in Political Science. An Introduction to Comparative Methods and Statistics*. Thousand Oaks: SAGE Publications Ltd.

Penrose, E. (1959). *The Theory of the Growth of the Firm*. New York: Wiley.

Pine, II, B. J. and Gilmore, J. H. (1998). Welcome to the experience economy. *Harvard Business Review*, 76(4), pp. 29–36.

Prahalad, C. K. and Hamel, G. (1990). The core competence of the corporation. *Harvard Business Review*, 68(3), pp. 79–91.

Richards, G. (2018). Cultural Tourism: A review of recent research and trends. *Journal of Hospitality and Tourism Management*, 36, pp. 12–21.

Rieznik, S. and Lee, H. (2018). The role of government in agricultural and rural development: Review of agricultural policies in Ukraine after independence with a look at the EU and South Korea experience. *Asian Journal of Agriculture and Rural Development*, DOI:10.18488/journal.1005/2018.8.2/1005.2.132.145

Rogers, E. M. (1995). *Diffusion of Innovations*. New York: Free Press.

Rogers, E. M. (2003). *Diffusion of Innovations.* New York: The Free Press.

Rogerson, C. M. and Rogerson, J. M. (2020). Camping tourism: A review of recent international scholarship. *Geo Journal of Tourism and Geosites*, 28(1), pp. 349–359.

Rosalina, P. D., Dupre, K. and Wang, Y. (2021). Rural tourism: A systematic literature review on definitions and challenges, *Journal of Hospitality and Tourism Management*, 47, pp. 134–149.

Sánchez-Zamora, P., Gallardo-Cobos, R. and Ceña-Delgado, F. (2014). Rural areas face the economic crisis: Analyzing the determinants of successful territorial dynamics. *Journal of Rural Studies*, 35, pp. 11–25.

Scowsill, D. (2011). *World Travel and Tourism Council Report.* Retrieved from https://www.onecaribbean.org/content/files/WTTCDavidScowsill.pdf (accessed: 10th August, 2021).

Sentosa, I., Sharin, F. H., Hadi, A. S. and Wahid, R. A. (2021). Ecotourism sustenance amidst the dynamic pandemic and recovery movement: The survival mechanism for creative entrepreneurs. In Proceeding Paper in 1st Warmadewa International Conference on Science, Technology, and Humanity "Ecotourism Resilience During Covid-19 Pandemic", Bali, 7th–8th September, 2021.

Sheena, B., Manohar, M. and Azlizam, A. (2015). Characteristics of Malaysian ecotourist segments in Kinabalu Park, Sabah. *Tourism Geographies*, 17(1), pp. 1–18.

Teece, J. D., Pisano, G. and Shuen, A. (1997). Dynamic capabilities and strategic management. *Strategic Management Journal*, 18(7), pp. 509–533.

Thahir, H., Hadi, S., Zahra, F., Arif, I., Murad, M. A. and Mashur H. Lolo (2021). Issues, Challenges and Strengths of Sustainable Tourism Supply Chain After Covid-19 in Togean National Park-Sulawesi, Indonesia: A Preliminary Findings. In Proceedings of the International Conference on Strategic Issues of Economics, Business and, Education (ICoSIEBE 2020). https://doi.org/10.2991/aebmr.k.210220.048

Thongdejsri, M. and Nitivattananon, V. (2019). Assessing impacts of implementing low-carbon tourism program for sustainable tourism in a world heritage city. *Tourism Review*, 74(2), pp. 216–234.

Tourism Malaysia (2015). *Tourism Malaysia Keen to Promote "Glamping".* Retrieved from: https://www.tourism.gov.my/media/view/tourism-malaysia-keen-to-promote-glamping (accessed: the 4th November, 2021).

United Nations (UN) (2015). *UN Sustainable Development Goals.* Retrieved from: https://sdgs.un.org/goals (accessed: the 5th November, 2021).

United Nations (UN) (2020a). *Entrepreneurship for Sustainable Development; Resolution adopted by the General Assembly.* Retrieved from: https://digitallibrary.un.org/record/3896576?ln=en (accessed: the 5th November, 2021).

United Nations (UN) (2020b). *United Nations Policy Brief: COVID-19 and Transforming Tourism.* New York: UN.

United Nations World Tourism Organization (UNWTO) (2020a). *UNWTO Launches Global Guidelines to Reopen Tourism.* Retrieved from: https://www.unwto.org/news/unwto-launches-global-guidelines-to-restart-tourism (accessed: the 4th November, 2021).

United Nations World Tourism Organization (UNWTO) (2020b). *World Tourism Barometer May 2020: Special Focus on the Impact of COVID-19.* Madrid: UNWTO.

Varley, P. and Semple, T. (2015). Nordic slow adventure. Explorations in time and nature. *Scandinavian Journal of Hospitality and Tourism*, 15(1/2), pp. 73–90.

Wallace, R. (2019). Ecotourism in Asia: How strong branding creates opportunity for local economies and the environment. In R. Hashim, M. H. M. Hanafiah and M. R. Jamaluddin (Eds.), *Positioning and Branding Tourism Destinations for Global Competitiveness*. Hershey: IGI Global, pp. 192–211.

Wall-Reinius, S. and Bäck, L. (2011). Changes in visitor demand: Inter-year comparisons of Swedish hikers' characteristics, preferences and experiences. *Scandinavian Journal of Hospitality and Tourism*, 11(1), pp. 38–53.

Weaver, D. B. and Lawton, L. J. (2002). Overnight ecotourist market segmentation in the Gold Coast Hinterland of Australia. *Journal of Travel Research*, 40(3), pp. 270–280.

Wong, J. W. C. and Lai, I. K. W. (2021). Effect of government enforcement actions on resident support for tourism recovery during the COVID-19 crisis in Macao, China. *Asia Pacific Journal of Tourism Research*, 26(9), pp. 973–987.

World Tourism Organization and Global Tourism Economy Research Centre (2017). *UNWTO/GTERC Annual Report on Tourism Trends, 2017 Edition: Executive Summary*. Retrieved from: https://www.wttc.org/economic-impact/country-analysis/country-reports/ (accessed: the 14th November, 2021).

9 Big Data Analytics in the Malaysian Aviation, Tourism, and Hospitality Industry

Rita Zaharah Wan-Chik and
Siti Salwa binti Hasbullah

Introduction

Big data addresses very large volumes of data which are automatically, and autonomously, generated or collected from various data sources, and it is characterized by the 5Vs: *volume*, *velocity*, *variety*, *veracity*, and *value*. Due to its nature of high volume and high speed, common challenges in big data include capturing, storing, searching, sharing, analyzing, and viewing the data, which requires new forms of data processing and analysis. Table 9.1 lists the corresponding challenges for each V in big data. Data analytics uses raw data to analyse the information gathered to uncover hidden patterns, correlations, and other insights. The different types of analytics offered by data analytics are descriptive, diagnostic, predictive, and prescriptive analytics. Descriptive analytics aid in the construction of the past pattern, diagnostic analytics aid in understanding why something happened in the past, predictive analytics aid in picturing a possible future, and prescriptive analytics advise on how to change such outcomes (Brinkman, 2020).

As technology advances, generating large amounts of data has become more cost effective due to low-cost storage, sensors, smart devices, and advances in computer multicore central processing units (CPUs), data storage, cloud computing, and open-source software (Akerkar, 2014). The adoption of this advanced technology in big data is not merely welcomed by computer professionals and database administrators, but is also being embraced by other players in various industries such as decision-makers, researchers, application users, and average consumers due to its ability to help companies improve their business decision-making, knowledge discovery, and process optimization (Akerkar, 2014). Big data is becoming cheaper and easier to manage as a result of new technology, and it is finding its way into more businesses and sectors, becoming a game changer in various industries (Analytics Insight, 2019). In the transportation industry, more governments and authorities are using big data in managing city congestion, transportation route planning, and intelligent transport systems as a whole, whereas in the private sector, big data is helping companies in revenue management, technological enhancements, and to gain a competitive advantage (Analytics Insight, 2019).

DOI: 10.4324/9781003315179-13

Table 9.1 5Vs characteristics and the corresponding challenges

Volume	Data in large volume, demands large storage capacity and analytical capabilities
Velocity	Demanding high velocity to process in real-time
Variety	Varieties of data types and formats
Veracity	Difficulty in tracing data or verifying data
Value	Varies with its application domains

(Source: Hwang and Chen, 2017.)

Data Analytics in Aviation, Tourism, and Hospitality

Aviation, tourism, and hospitality are global industries that play a vital role in connecting people, cultures, and businesses across the globe and that have also been taking advantage of this big data and data analytics solutions to better manage customer and transportation-related information in improving services and customer satisfaction. Massive amounts of data are being generated as more industries become digitally connected and reliant on computers and the Internet for easier operations as well as the communication of data and information.

Since the digital revolution began about a quarter-century ago, the aviation industry has been at the forefront of digital transformation. Companies all over the world are increasingly realizing that the key to unlocking the full potential of the aviation sector is not just to embrace digital transformation, but also to stay ahead of the curve by developing ground-breaking solutions based on new forms of digital technology. To drive the aviation business, as a process-oriented enterprise, the world of aviation relies on gathering, interpreting, analyzing, and monetizing data.

As a result of the evolution of sophisticated data analytics platforms and service providers, aviation companies and professionals may now observe a global transformation in the way data insights are used within the industry. Airline companies' approach to customer feedback is evolving with the dramatic increase in the level of feedback available from company websites and social media platforms and thus is improving user experiences. The data science revolution is revolutionizing nearly every part of the industry, which is continuing to generate more data from diverse sources and offering managers improved control and decision-making processes (Maire et al., 2017). Figure 9.1 shows the various data sources within the aviation industry. Consider how much data each data source generates on its own, and how much more if some of the data sources communicate and share information with one another. Data from these sources can be used for various data analytics such as weather forecasting, air pollution and congestion prediction, aircraft performance and maintenance prediction, and airspace congestion prediction, among others. There is so much more that can be done with the data to help organizations to make informed decisions and improve the future planning of fleet management, crowd management, and asset management.

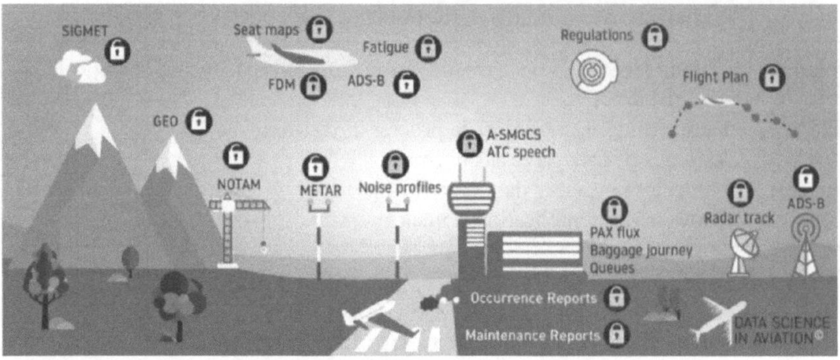

Figure 9.1 Various data sources in aviation (source: Fernandez, 2020).

In the tourism and hospitality industry, data analytics can help companies with revenue management by forecasting tourist demand and maximizing business revenues by combining internal data (e.g., current reservations and occupancy rates) with external data (e.g., information on upcoming events in the area, as well as school holidays and travel schedules). Market research and strategic marketing (e.g., analyzing customer patterns to best customize marketing opportunities), reputation management (e.g., online reviews, conversations on social media, service usage data, and customer feedback), and customer experience can all benefit from data analytics (Yallop and Seraphin, 2020).

Airports and Airline Customers

In the aviation industry, a huge amount of data is being generated by the airports and airline companies through their operation and commercial systems which may include data on flight operations as well as on the passengers. Flight passengers or airline customers' data when properly processed and analyzed using data analytics may provide a better understanding on customer or passenger behaviours. Airlines can benefit from the knowledge gained by analyzing a continuous stream of customer behaviour data to improve the entire passenger experience, such as offering personalized incentives for specific types of customers, which may help in better customer allegiance thus promoting greater percentages of repeat business and hence resulting in more auxiliary sales (Akerkar, 2014). By analyzing data collected from various platforms such as conversations in call centres, on-site forms, and social media, the use of data analytics may help companies to identify customers' intents and needs accurately. Companies can also look at the data behind consumer purchasing habits and click-through rates created on client application interfaces to develop a unified and responsive commercial strategy that can be adjusted in real-time as demand changes. Other than that, a dynamic study of competitors' pricing would enable airline companies and service suppliers to remain competitive at all times in the pricing war among airline companies, service providers, and travel agencies.

In the digital age, many airports across the world have introduced airport applications to help travellers access airport services in real-time via digital engagement, such as airport self-service kiosks, websites, or mobile applications for their travel needs. This includes flight bookings, check-ins, and checking for flight status, among others. Through these applications, airports can share with the travellers important information in real-time such as queuing time, flight status, gate updates, and time to gate, as well as estimated time before boarding. This is beneficial to rushed passengers since it allows them to avoid huge lines and keeps them informed about timetables and options from the moment they leave home for the airport (Maire et al., 2017).

The airport applications could both improve flight passengers' experiences and increase airlines' and airports' commercial revenues (Benitez, 2016). These applications can be designed as a marketing tool where flight passengers could be offered personalized products and services according to their needs or trends based on data generated using the application. Airport mobile applications can improve customer service by delivering real-time and accurate airport information, both within and outside the airport, such as flight data and updates, parking availability, ground transportation, and other airport facilities (Malaviya, 2015).

The application of Internet of things (IoT) and data analytics for airport operations covers a wide range of areas. It has provided many benefits as well as opening opportunities for airport operations. This is possible because IoT is a system of interconnected computing devices that can transfer data over a network without requiring human-to-human or human-to-computer interactions. Furthermore, data analytics uses raw data to create information for further use and multiple retrieval. This includes the usage of IoT cameras and sensors within the airport area to collect various types of data such as crowd movement and luggage movement. Beacon technology is also being utilized as part of IoT devices to collect data on airlines, baggage, and passengers for the purpose of planning service capacity, resources, and personnel (Kohli, 2016). One of the most important aspects in airport crowd management is to address, and thus reduce, long queues in favourite hotspots such as check-in counters, immigration checks, and boarding gates. Data analytics could be used on signals collected from passengers' cell phones when the passengers are within the airport's Wi-Fi sensors' range, predicting these hotspots. Data collected can be analyzed to see passengers' and crowd movement to help predict queuing or waiting times. Crowd behaviour and movement patterns could be used to predict future movements, based on historical data, which could assist in improving crowd management in the airport.

Maintaining a record of the huge volume of luggage that passes through airports every day is one of the most difficult challenges for airports and airline companies. More airlines and airports are adopting an improved luggage tracking and handling system with the use of radio-frequency identification (RFID) sensors as it has seen baggage handling improved by as much as 66 percent (Garcia, 2019).

RFID, consisting of a radio transmitter, a radio receiver, and a miniature radio transponder, uses electromagnetic fields to recognize and monitor tags affixed to items automatically. This would improve the luggage handling efficiency and promote a seamless experience for passengers.

Flights and Maintenance Management

An airplane alone, with the various systems and sensors fitted within its body, generates a huge amount of data. With the latest twentieth-century aircraft technology, it can produce 5 to 8 terabytes per flight, and by the year 2026, it is estimated that new aircraft will be producing up to 98 million terabytes of data (Maire et al., 2017). According to reports, data collected from one General Electric (GE) jet engine is at 5,000 data points per second, a Boeing 787 collects an average of 500 GB of system data per flight, and an Airbus A380 has up to 25,000 sensors to collect data from its aircraft (Shah, 2014). Much of this information is sent or downloaded in order to plan for maintenance, place order for spare parts, and predict component failure.

Data from the engine systems and fuel consumption, as well as other related flight operations, will be relayed in real-time from when the plane takes off until it touches down at its destination. Especially for the maintenance team, this data could assist them in making well-informed and on-time decisions on anticipated parts that would need replacement or repair. Analyzing data gathered from numerous sensors mounted to an aircraft could aid in predicting a component's failure (Akerkar, 2014). This could be used to evaluate the performance of the desired systems, including structural elements, electronics, hydraulics, and avionics as well as enabling real-time monitoring of corrosion (Ariffin et al., 2020), thus assisting in the decision-making process in optimizing the maintenance process.

The goal for an aircraft is to be able to offer full service for at least 18 hours a day for the next 15 to 20 years, and at the heart of this is optimizing the maintenance management of each aircraft. The price that airline companies pay for an aircraft for their fleet could be anywhere between 100 and390 million USD. Data analytics could offer predictive analysis for optimized maintenance management by using the massive amount of generated data from the aircraft as well as from relevant enterprise resource planning (ERP) systems providing logistic information used within the organization. The analysis may assist the management level in strategic maintenance planning and decision-making as well as assisting maintenance technicians to prepare the right parts and tools at the right time. Data analytics could aid in the monitoring of aircraft operations, the diagnosis of preventable malfunctions, and repair (Dou, 2020). This may lead the airline companies to optimize their operational efficiency, improving the company's operations with timely maintenance planning that can help to reduce operating costs and increase the investment in the aircraft (Badea et al., 2018).

Thus enabling early flight path modification if necessary, based on anticipated weather conditions. This will assist with the next iteration of aircraft by analyzing the structure information the aircraft will provide, thereby improving aircraft design and construction (Maire et al., 2017). The massive amount of data on aircraft movements generated by airport air traffic control centres might be utilized to safely manage air traffic, to construct more efficient flight routes, and to share data with airlines in order to increase flight performance and safety (Badea et al., 2018).

Tourism and Hospitality

Within the tourism and hospitality industries, a huge volume of data is generated from flight bookings and hotel bookings. Data analytics are used on this data to understand customer needs, behaviour, and satisfaction. Big data analytics and artificial intelligence (AI) solutions allow for the capturing of consumer preferences and the identification of market trends by taking into account other elements such as weather and local events that can influence visitor mobility (Marrese-Taylor et al., 2013; Analytics Insight, 2019). AI is defined as

> the ability of a digital computer or computer-controlled robot to perform tasks commonly associated with intelligent beings ... is frequently applied to the project of developing systems endowed with the intellectual processes characteristic of humans, such as the ability to reason, discover meaning, generalize, or learn from past experience.
>
> (Copeland, 2020)

Data analytics may be used to project actual and real-time forecasts of tourist movements, which can then be utilized to help create tourism industry plans (Samara et al., 2020). Data from the Web is the new way to learn about customer preferences for tourism items like hotels and restaurants. By combining price search and comparison, bookings, cancellations, and comments, a full picture of the client's interests and, more significantly, needs can be formed (Marrese-Taylor et al., 2013; Sygnal, 2017). In the tourism industry, data analytics could aid direct marketing methods by allowing a service to be tailored to the needs of the consumer such as offering an intelligent mechanism for personalized information to tourists depending on their preferences, feedback, and location. The use of big data analytics will increase supplementary revenue from secondary services such as partner company-specific agreements and a wide range of services such as customer browsing behaviour analysis. In the hospitality and tourism industry, the use of data analytics on data created in this industry helps tourism-related businesses personalize marketing, provide convenience to customers, save money, and gain a competitive advantage overall (Evans, 2020). It also aids in increasing sales income by offering vital business information for continuing marketing activity and advantage in the marketplace to hospitality and tourism organizations (Yallop and Seraphin, 2020).

Current Developments in Malaysia

It was reported in the Malaysia Artificial Intelligence Blueprint 2020 Annual Report that among the sectors in Malaysia, the early adopters of big data analytics are telecommunication, media and entertainment, aviation, and financial services (Bigittechnology, 2020). These industries are actively experimenting and deploying high-value data analytics and applying machine learning (ML) in various business cases and processes within their organizations to predict future outcomes that will assist their business strategic planning and operations. ML is a branch of AI that gives computers the ability to learn and grow from their experiences by learning from data supplied to the system. This allows the system to make decisions on its own, without the need for human interaction, using a computer algorithm based on what it has learnt (Expert.Ai, 2020). ML is a huge help for machine automation and automatic decision-making based on event predictions.

One of the world's major airport operator groups, Malaysia Airports Holding Berhad (MAHB), which handles 39 airports in Malaysia, including the Kuala Lumpur International Airport (KLIA) and its second terminal, the Kuala Lumpur International Airport 2 (KLIA2), has begun its digital transformations in those airports, which include the use of IoT devices and big data analytics to improve airport operations, with the ultimate goal of improving passenger comfort and lowering queue time and airport congestion (International Airport Review, 2018). Data management is used in its integrated software suite on a full-fledged analytics platform that would deliver predictive analytics for its service management and asset management to ensure a seamless customer experience while also serving the larger goal of business and operational excellence (Bansal, 2018). Additionally, this digital transformation will also offer benefits to MAHB's commercial partners, airport retailers as well as the immigration department operating in the airports. By combining the power of IoT and ML, these solutions will strengthen MAHB predictive maintenance capabilities, thus supporting more precise decision-making. More recently, MAHB has migrated its network to a new core network architecture at KLIA, allowing the airport to take advantage of cutting-edge technologies such as 5G, WiFi6, and IoT, as well as implementing the Airport Collaborative Decision Making (ACDM) system to help the airport manage its resources more effectively and efficiently, particularly in-flight operations (Focus Malaysia, 2020).

MYairports Mobile Application

MAHB has released the MYairports mobile application (MYairports app) that is available to be used at both KLIA and KLIA2 to help travellers to plan their journey at their fingertips. The launch of this application represents Malaysia Airports' progress toward embracing Airports 4.0, in which real-time digital engagement with passengers is critical and considered a must have (Malaysia Airports, 2018). With the help of data analytics, this application can assist travellers to get real-time updates on their flight schedules, estimated time left to board their flight, shopping and dining promotions, and important airport information. Travellers will receive reminders of their flights as well as other vital items to bring before departing, such as their passport, boarding pass, flight check-in, travel visa, and wallet, by entering their flight number or scanning their boarding card into the application. Users of the application will also receive options on the fastest routes to get to the airport, and once in the airport, they can use the indoor navigator to bring them to the desired spots within the airport by the fastest route using the dynamic way-finding feature in the MYairports app. This way-finding feature is the most crucial feature in this application because indoor navigation could help travellers navigate the airport efficiently (Kamaruddin, 2015). Additionally, travellers will be able to see a countdown of time left to check-in and board their flights. The main goal of the MYairports app is to help passengers navigate their way from home to the boarding gate in the shortest time possible. More technological developments are being developed to improve the airport operations by forecasting traffic volumes, managing passenger facilities, and decreasing airport queuing time and congestion (Malaysia Airports, 2018).

Single-Token Biometric ID

Biometrics is the measuring and statistical analysis of the physical and behavioural characteristics of a person. It is mostly used for identification, access control, and individual identification. Biometrics operates by collecting data from the passengers' various biometrics, which is then processed by the biometrics system. The biometric system would be able to identify passengers based on the specified and unique data acquired by the various types of biometrics. Physiological and behavioural biometric identifiers are the two forms of biometric IDs. The data of the travellers is obtained using these identifiers. Fingerprints, as well as the location and size of the finger, face recognition, the eye, such as the iris and retina, voice, and DNA fall under the physiological category, whereas behaviour includes typing patterns, bodily movements, and other gestures (Korolov, 2019). As part of the Airports 4.0 plan, which intends to leverage big data analytics to transform KLIA into a smart airport, single-token biometric identification has been introduced that no longer requires the checking of physical identity at multiple checkpoints in the airport (Malaysia Airports, 2020). The Airports 4.0 concept is a metamorphosis into a smart airport, which uses single-token data to let passengers effortlessly navigate all touchpoints and provide personalized services. Passengers will be required to scan their faces at every touchpoint to validate their "token," and will not be required to submit their passports or boarding cards at many checkpoints throughout the airport, with the token thus serving as the passenger's new passport (Sinha, 2019). All relevant parties, such as airlines, immigration, and customs, will be given access to the information. Because the passenger's face will serve as both a boarding permit and a passport, passengers will no longer be required to provide their passports or boarding cards at different checkpoints throughout the airport. This approach is automating the verification processes, helping in the smooth process of passengers' movement in the airport as well as reducing the burden of airport staffing as fewer passenger verification processes are needed (Sinha, 2019).

Future Developments in Malaysia

As for future and continuous development to help further improve the aviation, tourism, and hospitality industry, several technologies, either supporting or using data analytics, which could be adopted are discussed below.

Beacon Technology

Beacons are utilized as part of IoT technology to provide information on passengers, airlines, and baggage that can help airports plan service capacity, resources, and staffing more effectively. KLIA can employ beacon technology in combination with the MYairports app to provide real-time "day of travel" information to travellers in the airport, such as gate changes or flight delays. Beacons can assist travellers connecting to another aircraft by allowing the airport to send directions directly to the traveller's smartphone. The airport may send directions to the next aircraft to the traveller promptly, skipping the packed information kiosk. Arriving travellers could receive a message from a beacon in baggage reclaim, for example, advising them of which carousel their luggage will arrive on and how long they will have to wait. If KLIA sees a high number of travellers pinging a beacon near security checkpoint 3, but a low volume of travellers pinging the beacon near checkpoint 1, resources can be diverted to either move passengers to the less busy security line. "or to move passengers to the less crowded security line" can be removed. Alternatively, more personnel might be dispatched to the busy spot. This not only enhances the airport's and visitors' experiences, but the data can also be utilized to create predictive models that help the KLIA better plan for and distribute resources where they are most needed before an issue occurs.

The Airports Council International (ACI) and International Air Transportation Association (IATA) have collaborated with aviation, airports, airline, and technology industry experts to develop the "Airport Terminal Beacons Recommended Practice" (Airport Council International, 2018). The placement and use of airport terminal beacons is outlined in this Recommended Practice. It also includes technical and managerial knowledge needed to configure, operate, and maintain airport beacons across the terminal, as well as information on how to deliver a beacon service. KLIA can follow this Recommended Practice to successfully implement beacon infrastructure and deliver beacon services at terminals and could improve further features in the current MYairports app.

Air Traffic and Airside Management

The Department of Civil Aviation (DCA), within the Malaysian Ministry of Transport, operates the Air Traffic Control System (ATCS) in KLIA. It was first commissioned in 1998 and has since assisted DCA KLIA in the safe and efficient transit of aircraft to and from the airport, as well as meeting the department's operational availability and functionality standards. The Display Processing Surveillance Data

(DPSD) is now used by the ATCS in KLIA to show, track, monitor, and identify the movement of arriving and leaving aircraft in the vicinity of the airfield, and on the runways and taxiways, as well as vehicles working in the manoeuvring zones (Civil Aviation Authority of Malaysia, 2020). Radar data from various radar systems, weather data from various weather systems, runway incursions and stop-bar status from above ground level (AGL), flight plan data, and navigational aids status from the Flight Data Processing System (FDPS) are all processed and displayed by the DPSD. These systems generate huge amounts of data but unfortunately, some of these systems are standalone and do not share information with each other. To support a more efficient airport and flight operations monitoring, an integrated system that connects all relevant systems, equipped with data analytics capability, could enormously help to improve airport operations monitoring and performance. The integrated system could assist in catering for the ever-increasing expectations for more efficient traffic flows and airport operations. It may also help to increase the capacities of limited airside and airport resources like runways and airspace as well as helping to make airline operations more economical and environmentally responsible by reducing fuel consumption and greenhouse gas emissions by having more efficient air traffic and airside management integrated systems.

Big Social Data

Data created by airlines and tourism social media could be used and analyzed to assist relevant enterprises in improving their services in order to increase consumer satisfaction and loyalty while remaining competitive with rival companies. Companies might expand the quantity of data they collect from social media, allowing them to give better service to clients, keep and develop them, and grow their business (Chen et al., 2016). Travel expenses and feedback data could be analyzed using data analytics to establish a comprehensive traveller profile, and companies could improve flight arrangements and services accordingly which could help increase customer satisfaction, retention, and loyalty. Personal marketing campaigns could also be tailored based on these established profiles to help improve the marketing conversion rate. The value and behaviour of other users within the same social relationship are influenced by the link between social media users, so social media data could be used for viral marketing based on a customer's social graph. Sentiment analysis of comments on various social media sites or Twitter retweets can be used to gain a better understanding of how these individuals interact and how their value translates to greater sales and revenue. Future research can combine passenger demographic profiles with other offline behaviour data to better estimate the traveller's overall value to airlines and tourism organizations (Chen et al., 2016).

Conclusion

It has been proven that the upgrade in technology and more wide-spread use of information technology in the aviation industry have contributed to improvement in operational efficiency and safety, as well as customer satisfaction (Rencher et al., 2012), and it is predicted that with the use of data analytics things are going to improve in more ways than one can imagine. Big data analytics plays a critical part in the development of smart aviation as well as increasing an aircraft's safety and performance (Dou, 2020). It can improve aircraft safety and performance by creating more effective predictive models (Douglas, 2014). Efficiencies that smooth out operations could help result in fewer travel delays and inconveniences, and these are important improvements that help keep customers happy (Maire et al., 2017). With better aircraft design and performance enhancement, it is possible to increase the technical performance and operating circumstances, as well as avoiding aircraft malfunctions (Dou, 2020), which would make planes easier to fly and maintain while also increasing their efficiency (Maire et al., 2017). For strategic revenue management, intelligent price management can solve the difficulty of setting the proper price in relation to time across a variety of distribution channels for tourist items (e.g., airplane tickets, hotel rooms) (Noh et al., 2016).

The following are other important and major benefits of applying data analytics for aviation, tourism, and hospitality in the current state of development:

- Offers predictive capabilities to promote aviation safety, making planes safer and more reliable (Akerkar, 2014; Barak and Dahooei, 2018; Ni et al., 2019; Dou, 2020)
- Improves customer-centred planning by assisting airline company decision-makers in better identifying problems and mining deeper knowledge (Dou, 2020)
- Helps airports with designing routes, air traffic management, air logistics management, flight management including environment and safety, airport and service management, crew management, and other aviation operations (Dou, 2020)
- Provides effective business solutions for the aviation and tourist industries, and uses data to improve service levels and client retention (Akerkar, 2014; Kim, 2016; Park and Pan, 2018)
- Helps to identify the elements that attract consumer retention and satisfaction to predict tourism loyalty (Hsu et al., 2009)
- Provides a deeper knowledge of travellers' diverse preferences for environmentally friendly destinations (Lopez-Sanchez and Pulido-Fernández, 2017)

However, in realizing many of these potential offerings and efficiencies from data analytics, several obstacles are found to hinder the success of adopting the full capabilities big data and data analytics can offer. Some of these challenges include data incompatibility between systems and technologies. Data from legacy information technology and systems is sometimes found not to be flexible enough to work with the latest and more sophisticated data analytics systems. The conversion or transformation of data from legacy systems to the latest data analytics systems is also expensive and time consuming. Another challenge is the lack of data scientists and engineers able to operate the sophisticated data analytics systems and understand the more advanced process and its output. Specialized forecasting agencies and staff will be required, particularly for aviation forecasts involving aeronautical meteorological predictions, and high-altitude forecasts (Dou, 2020).

According to the Malaysia AI Blueprint 2020 Annual Report, Malaysia is currently at Level 3 (Systematic) in its usage of big data and data analytics, with organizations and enterprises using it on specific projects to predict the future and improve current operations (Bigittechnology, 2020). Some organizations are using ML to predict future outcomes in various business cases, while others are currently developing AI-driven products and solutions based on successful proofs-of-concept (POC). Descriptive analytics, the most basic form of data analytics, still dominate projects, delivering a summary of what has happened by exploiting historical data.

There are a number of obstacles to overcome, specifically in Malaysia, in order to successfully implement big data analytics. The scarcity of top-notch data science specialists, a lack of a cohesive business plan, and insufficient cash to drive advanced analytics initiatives are at the top of this list of adoption hurdles (Bigittechnology, 2020). To help Malaysian organizations to attain successful big data analytics and AI maturity, these are several recommendations from the report (Bigittechnology, 2020):

- Upskilling – invest in continuous upskilling and training programmes in advanced analytics in predictive analytics (to forecast future events), data engineering (design, manage and optimize data pipelines), and big data analytics strategy (laying out a comprehensive data-driven roadmap).
- Clear strategy – better objectives for the effective execution of data initiatives, clear company vision and a data-driven roadmap, merging advanced analytics into existing business processes.
- C-level mandate – support and advocacy from management and stakeholders to ensure a thriving data-driven culture and adoption.

In conclusion, for the aviation industry, data analytics adoption is helping the industry in airline and airport performance management, weather forecasting, air pollution and congestion prediction, aircraft performance and maintenance prediction, and airspace congestion prediction, and improving the future planning of fleet management, crowd management, and asset management. The adoption of data analytics on data collected within the tourism and hospitality industry helps to facilitate personalized marketing and customer convenience, save expenses, and gain advantage in the marketplace. As discussed in this chapter, organizations and businesses could benefit so much from the adoption of data analytics in their operations and strategic decision-making. However, to optimize and gain data analytics' fullest capabilities will need comprehensive and sufficient information generated from various data sources and accurate prediction systems operated by skilled data scientists and engineers.

References

Airport Council International (ACI) (2018). *Airport Terminal Beacons Recommended Practice.* Retrieved from: https://aci.aero/Media/6b4d6043-1bab-457c-8352-53c4c51dd634/5XBxhA/About%20ACI/Priorities/IT%20-%20New/Initiativies/Beacons/ACI-Airport-Terminal-Beacons-Recommended-Practice.pdf (accessed: the 25th July, 2021).

Akerkar, R. (2014). Analytics on big aviation data: Turning data into insights. *International Journal of Computer Science & Applications*, 11(3), pp. 116–127.

Analytics Insight (2019). *10 Industries Redefined by Big Data Analytics.* Retrieved from: https://www.analyticsinsight.net/10-industries-redefined-by-big-data-analytics/ (accessed: the 15th July, 2021).

Ariffin, A. H., Mahzan, S. and Mohamad, Z. (2020). A brief review on structural health monitoring sensor technology in civil and aviation technology applications. *Journal of Advanced Mechanical Engineering Applications*, 1(2), pp. 1–6.

Badea, V. E., Zamfiroiu, A. and Boncea, R. (2018). Big data in the aerospace industry. *Informatica Economica*, 22(1), pp. 17–24.

Bansal, G. (2018). Malaysian Airport integrates data analytics to enhance customer satisfaction. *Analytics Insight.* Retrieved from: https://www.analyticsinsight.net/malaysian-airport-embraces-data-analytics-to-deliver-world-class-customer-satisfaction/ (accessed: the 18th July, 2021).

Barak, S. and Dahooei, J. H. (2018). A novel hybrid fuzzy DEA-Fuzzy MADM method for airlines safety evaluation. *Journal of Air Transport Management*, 73, pp. 134–149.

Bigittechnology (2020). *Malaysia AI Blueprint 2020 Annual Report.* Retrieved from: https://www.bigittechnology.com/malaysia-ai-blueprint-2020/ (accessed: the 18th July, 2021).

Benítez, L. F. (2016). The image of the airport through mobile applications. *International Journal of Scientific Research in Information Systems and Engineering (IJSRISE)*, 2(1), pp 58–62. Retrieved from: https://www.researchgate.net/publication/302872614_The_image_of_the_airport_through_mobile_applications (accessed: the 18th July, 2021).

Brinkman, B. (2020). Comparing descriptive, predictive, prescriptive, and diagnostic analytics. *LogiAnalytics.* Retrieved from: https://www.logianalytics.com/predictive-analytics/comparing-descriptive-predictive-prescriptive-and-diagnostic-analytics/ (accessed: the 18th July, 2021).

Chen, S., Huang, Y. and Huang, W. (2016). Big data analytics on aviation social media: The case of china southern airlines on sina weibo. In 2016 IEEE Second International Conference on Big Data Computing Service and Applications (BigDataService) (pp. 152–155). Oxford, UK: IEEE. 29th March–1st April, 2016.

Civil Aviation Authority of Malaysia (CAAM) (2020). *Air Traffic Control.* Retrieved from: https://www.caam.gov.my/resources/aviation-professional/air-traffic-control-atc/ (accessed: the 3rd July, 2021.

Copeland, B. J. (2020). Artificial intelligence. *Encyclopedia Britannica.* Retrieved from: https://www.britannica.com/technology/artificial-intelligence (accessed: the 5th July, 2021).

Dou, X. (2020). Big data and smart aviation information management system. *Cogent Business & Management*, 7(1), p. 1766736.

Douglas, C. C. (2014). An open framework for dynamic big-data-driven application systems (DBDDAS) development. *Procedia Computer Science*, 29, pp. 1246–1255.

Evans, N. (2020). *Strategic Management for Tourism, Hospitality and Events.* Abingdon: Routledge.

Expert.Ai (2020). *What is Machine Learning? A Definition.* Retrieved from: https://www.expert.ai/blog/machine-learning-definition/#:~:text=Machine%20learning%20is%20an%20application,it%20to%20learn%20for%20themselves (accessed: the 5th July, 2021).

Fernandez, A. (2020). Aviation data compendium: Which data sources exist in aviation industry? *DataScience.Aero.* Retrieved from: https://datascience.aero/data-sources-aviation/ (accessed: the 5th July, 2021).

Focus Malaysia (2020). *MAHB Aspires KLIA to be Digital Airport.* Retrieved from: https://focusmalaysia.my/category/featured/ (accessed: the 10th July, 2021).

Garcia, M. (2019). Airlines are losing fewer bags thanks to baggage tracking technology. *Travel+Leisure.* Retrieved from: https://www.travelandleisure.com/airlines-airports/airlines-baggage-tracking-technology-improvements (accessed: the 18th July, 2021).

Hsu, C. I., Shih, M. L., Huang, B. W., Lin, B. Y. and Lin, C. N. (2009). Predicting tourism loyalty using an integrated Bayesian network mechanism. *Expert Systems with Applications*, 36(9), pp. 11760–11763.

Hwang, K. and Chen, M. (2017). *Big-data Analytics for Cloud, IoT and Cognitive Computing.* Chichester: John Wiley & Sons.

International Airport Review (2018). *Malaysia Airports Launches New Passenger Experience App based on Big Data Analytics.* Retrieved from: https://www.internationalairportreview.com/news/70116/malaysia-airports-app/ (accessed: the 10th July, 2021).

Kamaruddin, M. A. (2015). Low cost indoor navigation system for KLIA2 using mobile application. IRC. Retrieved from: http://utpedia.utp.edu.my/15842/2/Dissertation%20FYP2%20Report.pdf34.pdf (accessed: the 15th July, 2021).

Kim, S. (2016). Forecasting short-term air passenger demand using big data from search engine queries. *Automation in Construction*, 70, pp. 98–108.

Kohli, D. (2016). Four ways beacon technology can benefit airports and travellers. *Airport Focus International.* Retrieved from: http://airportfocusinternational.com/four-ways-beacon-technology-can-benefit-airports-and-travellers/ (accessed: the 26th July, 2021).

Korolov, M. (2019). What is biometrics? 10 Physical and behavioral identifiers that can be used for authentication. *CSO ASEAN.* Retrieved from: https://www.csoonline.com/article/3339565/what-is-biometrics-and-why-collecting-biometric-data-is-risky.html (accessed: the 10th July, 2021).

Lopez-Sanchez, Y. and Pulido-Fernández, J. I. (2017). Factors influencing the willingness to pay for sustainable tourism: A case of mass tourism destinations. *International Journal of Sustainable Development and World Ecology*, 24(3), pp. 262–275.

Maire, S., Krishnan, V. and Spafford, C. (2017). Aviation's data science revolution. *Oliver Wyman.* Retrieved from: https://www.oliverwyman.com/our-expertise/insights/2017/jun/aviation-s-data-science-revolution.html (accessed: the 15th July, 2021).

Malaviya, A. (2015). Airports move to mobile. *Aviation Pros.* Retrieved from https://www.aviationpros.com/engines-components/aircraft-airframe-accessories/cabin

-communications/article/12039240/airport-mobile-technology (accessed: the 20th July, 2021).

Malaysia Airports (2018). *Enjoy a Whole New Digital Airport Experience with MYairports, the New Mobile App by Malaysia Airports.* Retrieved from: https://www.malaysiaairports.com.my/media-centre/news/enjoy-whole-new-digital-airport-experience-myairports-new-mobile-app-malaysia (accessed: the 21th July, 2021).

Malaysia Airports (2020). *KLIA Pilots Facial Recognition Technology Throughout the Total Passenger Journey.* Retrieved from: https://www.malaysiaairports.com.my/media-centre/news/klia-pilots-facial-recognition-technology-throughout-total-passenger-journey (accessed: the 21st July, 2021).

Marrese-Taylor, E., Velásquez, J. D. and Bravo-Marquez, F. (2013). Opinion zoom: A modular tool to explore tourism opinions on the web. In 2013 IEEE/WIC/ACM International Joint Conferences on Web Intelligence (WI) and Intelligent Agent Technologies (IAT) (pp. 261–264). Atlanta, GA, USA: IEEE. 17th November-20th November, 2013.

Ni, X. M., Wang, H. W., Che, C. C., Hong, J. Y. and Sun, Z. D. (2019). Civil aviation safety evaluation based on deep belief network and principal component analysis. *Safety Science*, 112, pp. 90–95.

Noh, S. Lee, H.-C. and Lee, S. K. (2016). *Exploring the Determinants of Strategic Revenue Management with Idiosyncratic Room Rate Variations.* Retrieved from: https://agrilifecdn.tamu.edu/ertr/files/2016/12/RN99.pdf (accessed: the 18th July, 2021).

Park, S. Y. and Pan, B. (2018). Identifying the next non-stop flying market with a big data approach. *Tourism Management*, 66, pp. 411–421.

Rencher, R., Whitlock, S. and Francy, F. (2012). Securing airline information on the ground and in the air. *AERO.* Retrieved from: https://www.boeing.com/commercial/aeromagazine/articles/2012_q3/5/ (accessed: the 24th July, 2021).

Samara, D., Magnisalis, I. and Peristeras, V. (2020). Artificial intelligence and big data in tourism: A systematic literature review. *Journal of Hospitality and Tourism Technology*, 11(2), pp. 343–367.

Shah, D. (2014). Millions of data points flying in tight formation. *Aerospace Manufacturing and Design.* Retrieved from: https://www.aerospacemanufacturinganddesign.com/article/millions-of-data-points-flying-part2-121914/#:~:text=For%20the%20past%20several%20years,as%20many%20as%2025%2C000%20sensors (accessed: the 25th July, 2021).

Sinha, R. (2019). IOT in aviation industry. *ESDS.* Retrieved from: https://www.esds.co.in/blog/iot-in-aviation-industry/#sthash.6anYJhcm.dpbs (accessed: the 18th July, 2021).

Sygnal (2017). *How is Big Data in Aviation Transforming the Industry?* Retrieved from: https://www.we-are-sygnal.com/how-is-big-data-in-aviation-transforming-industry/ (accessed: the 24th July, 2021).

Yallop, A. and Seraphin, H. (2020). Big data and analytics in tourism and hospitality: Opportunities and risks. *Journal of Tourism Futures*, 6(3), pp. 257–262.

10 An Empirical Study on the Technology Usage Dimensions within the Tourism Craft Industry in Bali, Indonesia

A Structural Equation Modelling Approach

Putu Ngurah Suyatna Yasa and Ilham Sentosa

Introduction

The importance of the role of small and medium enterprises (SMEs) as a supporter of Bali tourism has been recognized by the governor of Bali through the regional development vision for 2019–2024. It is closely related to the empowerment of Balinese local wisdom, known as "Nangun Sat Kertih Loka Bali Through the Development of the Planned Universe." It is also related to the development of SMEs in Bali. The vision is then also stated in the Mission as: Mission 1: Ensuring the fulfilment of food, clothing, and shelter needs in sufficient quantity and quality for the life of the Balinese. Mission 2: Realizing food self-sufficiency, increasing the added value and competitiveness of agriculture, and improving farmers' welfare. Mission 6: Developing highly competitive human resources, namely quality and integrity, professionalism, and morality that are developed based on the values of the local wisdom of the Balinese. Mission 8: Producing a competent, productive, qualified, and highly competitive workforce and expanding access to job opportunities at home and abroad. Mission 13: Developing new tourism destinations and products based on culture and in favour of the people that are integrated among regions/cities throughout Bali. Mission 16: Building and developing new economic centres in accordance with the potential of districts/cities in Bali by empowering local resources to support economic growth in a broad sense. And Mission 17: Building and developing small and medium-sized industries based on culture (Bali branding) to strengthen the economy of the Balinese.

The substance of the vision and Mission is to empower the Balinese people's economy by treating SMEs as an important supporting sector in increasing economic growth. The importance of the role of SMEs can be seen in Bali's gross domestic product (GDP), where the processing industry, with 16 types of business fields including SMEs, contributed to Bali's GDP on average about 6% per year during the period 2014–2018 (Central Bureau of Statistics, 2019) or occupied the position of the 6th largest contributor of all business fields.

DOI: 10.4324/9781003315179-14

Likewise, manufacturing/processing industries, including SMEs, occupy the second highest position in the province of Bali after wholesale and retail trade.

The number of SMEs in Bali as of August 2018 increased by 17% from 2016 with added value growth in 2017 of 17% (Director General of Indonesian Small and Medium Industries, 2018). The growth in the number of SMEs in Bali is inseparable from the increasing number of tourist visits, both foreign and domestic tourists. In addition, it also shows the increasing interest in entrepreneurship among the younger generation which currently ranks no. 2 in Indonesia after Yogyakarta. SMEs also have a strategic role in absorbing workers in Bali Province, and are in first position compared to 17 other business fields (Central Bureau of Statistics, 2016).

One part of the growing SMEs in Bali is the small handicraft industry (SHI); its role in the Balinese economy is as one of the key export commodity sectors in addition to livestock, textiles, garments, and paper. The SHI in this study included non-machine weaving, batik, lace, Balinese dance clothing, manuscripts, wood sculpture carvings, iron crafts, bamboo crafts, gold/silver crafts, painting, pandanus weaving, painted egg crafts, and barong crafts. SHI's contribution to Bali's gross regional domestic product (GRDP) increased from 1.82% in 2006 to 1.88% in 2009. Likewise, the value of handicraft exports to Bali's total exports is the second largest after apparel, at 13.94% (Central Bureau of Statistics, 2010).

A report by Bank Indonesia (2010) shows that SHI in Bali has difficulty in increasing its product and business scale. Theoretically, efforts to increase business productivity are closely related to the use of a combination of production factors used in the production process. In traditional neoclassical growth theory, growth is always related to the use of the three factors of production, namely capital, labour, and technology. The theory was further refined by Romer (1986) in the New Growth Theory or the so-called endogenous growth model which is the basis of this research theory. This view develops among developing countries, giving rise to the view that technology is an endogenous factor that can be influenced by various policy variables. The source of growth in this theory is the increasing stock of knowledge and new ideas in the economy which encourages the growth of creativity and initiative in innovative and productive activities that demand increased quality of human resources. The transformation of knowledge and new ideas can occur through international trade, investment, licensing, consulting, and communication activities (Romer, 1986).

Technology increases efficiency in production, namely an increase in output produced while using fewer inputs, or an increase in output with fixed inputs. In general, the higher the technology content, the greater the investment required in the production process. The utilization of technology can be neutral, namely increasing output without having to add input factors, or a bias towards the use of production factors, so that technological progress does not always increase the use of labour (labour saving) or save the use of capital (Simangunsong et al., 2019).

In SHI, the types of technology used include traditional technology, modern technology, and/or a combination of the two, called mixed technology.

Traditional technology includes all types of work tools that are driven by hand without using mechanical power, electricity, computers, or batteries, such as carving tools (chisels and chisels), non-machine wood cutting tools (saws), and traditional wood smoothing tools (shavings), pattern-making equipment, and paint equipment. Modern technology is any type of equipment that is driven by engine power, electricity, computers, or batteries (Nala, 1989). Modern technology is widely used in the production process, for communication (telephone, facsimile, and internet), and transportation. The type of work technology used today is generally a mixture of traditional and modern technology, because not all types of production activities can be carried out using modern technology, especially activities related to artistic values. Likewise, modern technology plays a very important role in increasing productivity, especially for export-based activities or relatively large-scale businesses.

The importance of the role of technology in increasing business growth can be proven from the results of research: Kuznets in 1966 in the United States (Amir, 2008) showed that the contribution of technological progress to economic growth in 1929–1957 was 78%, and in the period 1950–1962 it was 56%. The Solow study (Baier et al., 2006) said that the growth of the American economy between 1900 and 1949, by 80%, was triggered by the role of technology which was included in the total factor productivity. Levy's study, as in Husni (1994), shows that industrial growth in Iraq between 1961 and 1967 was labour saving due to advances in production technology. Restuccia's study shows that the use of modern technology in SMEs can reduce production barriers compared to traditional technology (Restuccia, 2004). The results of the 1972 Field study in South Korea in 1961–1980 (Husni, 1994) showed that industrial growth tends to be capital-intensive triggered by the use of technology.

Amir (2008) mentioned that in five developed countries, namely France, Japan, West Germany, England, and the United States of America, technology contributed to economic growth by 49–76%. According to a Japanese study on the steel industry, technological change will decrease company productivity for the first two years, but after that, it will grow significantly more than the 4.10 percent reduction that occurred at the start of the shift in technology usage (Nakamura, 1981). Previous research in the 1960s showed the role of technological change in increasing the GDP by 0.32% in the Japanese economy (Nakamura, 1981). Again, Nakamura (1981) shows that companies that are less productive generally lack support for the use of technology in production. Subsequent studies show that increased business productivity can be realized through increased technology investment (Thatcher and Oliver, 2001).

Kerr and Newell (2003) show that technological changes must be made for companies that want to exist and still be able to compete. Indonesia's economic growth in the five years from 2000 to 2004 was triggered by capital and modern technology (Wijono, 2005). Likewise, the results of Prabawa and Budhi's research (2017) on the screen printing industry in the city of Denpasar found that capital, wage levels, technology, and employment had a positive and significant effect on productivity. Furthermore, the study of Duan, Mullins, Hamblin, Stanek, Sroka,

Machado, and Araujo (2002) shows that the use of IT in various organizations is able to reduce transaction costs and overcome distance barriers in coordination, which in turn supports the potential for coordination in SMEs at every stage of the business process. The use of IT in organizations including SMEs has the potential to reduce costs and increase business productivity. Ladokun and Osunwole (2013) also stated that information technology (IT) used in communication and collaboration management, customer access, managerial policy making, data management, and knowledge management increases organizational productivity and services.

On the other hand, the studies show that technology does not always have a positive and significant influence on business growth. The results of a study in Indonesia using data from 1969 to 1993 show that technology's contribution is only 4.01% against average annual economic growth of 7.11% (Amir, 2008). The study of Bosworth et al. in Indonesia in 1995 using data from 1964 to 1980 (Amir, 2008) shows that the contribution of technology to supporting output growth is 19.45%. Amir (2008) shows that technology was not the main variable in increasing business growth, but human resources are more important than technology; technology cannot work without human resources managing the technology, because humans have a role in regulating technological production factors. There are conflicting results from these research results, so to ensure the role of technology in business growth in Bali, it is necessary to conduct research on the role of technology in SHI as a tourism supporter.

SMEs in Indonesia make up 99% of all business activities, providing 89% of private sector employment and absorbing 65% of the total workforce in the industrial sector and contributing 60% of GDP (Asia Pacific Foundation of Canada, 2018). However, behind the relatively large contribution, especially in the absorption of labour, SMEs in general have relatively low productivity. Overall, the value added ratio per worker for small businesses is only about 1/200 times the value added per worker for large businesses; this low productivity is a major obstacle in the development of SME businesses in Indonesia. The productivity of SMEs which is still low turns out to be positively correlated with the productivity of Indonesian workers in general, which is nationally ranked number five out of ten Association of Southeast Asian Nations (ASEAN) countries (ASIAN Productivity Organization, 2019). In addition, low productivity is also influenced by the factor of technology utilization which is also relatively low, both for production technology and information technology.

The productivity of SMEs in Bali is not different from conditions at the national level; SMEs in Bali generally have low productivity compared to large industries. In the last five years the Bali Regional Government has tried to encourage the performance of SMEs; in the second quarter of 2017 SMEs in Bali experienced growth of 3.87%, higher than the national figure of 1.32% (Central Bureau of Statistics, 2018). However, this growth was much smaller than the growth of the SMEs, which was 8.95%. Based on this description, it is very important to do research on the impact of using technology on the growth of SHI in Bali.

Literature Review

Defining Economic Growth

Economic growth is a benchmark which is commonly used to determine the economic performance of a country, describing the increase in total output, known as GDP, in the long term. In traditional neoclassical growth theory as the basis for the theory of endogenous growth from Romer (Todaro and Smith, 2006), that output growth always comes from one or more of the three production factors, namely capital production factors through increased savings and investment, labour through increasing the quantity and quality of manpower, and improving technology.

Neoclassical theory views the role of technology as important because it is able to increase production efficiency through three basic classifications, namely: (1) neutral technological progress (i.e., a higher level of output is obtained with the same quantity of input combinations); (2) labour-saving technological progress, namely technological progress that can save the use of labour production factors; (3) capital-saving technological progress that occurs in many developing countries. The use of more advanced technology can increase labour productivity.

Technological advances can cause an increase in the marginal productivity of labour (MPL) and decrease the marginal productivity of capital (MPK), so as to save the use of capital production factors/capital saving, or reduce MPL and increase MPK so as to save labour (Allen, 1967). This condition will occur with the assumption that factor prices remain relatively unchanged (Layard and Walters, 1978). Meanwhile, production activities that do not experience change/ that are neutral will occur if the production process does not cause the rise or fall of MPL and MPK.

The theory of endogenous growth pioneered by Romer (1986) is the beginning of a revival and understanding of the factors that determine economic growth in the long run. Along with the development of the world marked by the development of modern technology in various production activities, the problems of economic growth such as decreasing returns to capital, perfect competition, cannot be explained properly by neoclassical theory. Endogenous growth theory has three principal elements, namely: (1) endogenous technological changes through the process of knowledge accumulation, (2) the creation of new ideas by companies as a result of the spillover mechanism and learning by doing, and (3) the production of consumer goods is supported by knowledge that grows without limits. Endogenous growth theory explains that growth in the long term is determined from within the model by several growth variables that were previously considered exogenous (Romer, 1986; Barro and Sala-i-Martin, 1999).

Furthermore, Romer's (1986) theory emerged as a critique of neoclassical growth theory regarding the decline in the marginal productivity of capital and the convergence of income. However, based on empirical studies, it shows that there is no convergence of income in various countries (Romer, 1986). This is because developed countries have developed technologies that can increase their production capacity. Technological progress is supported by the availability of quality human resources, so that they can carry out technological innovations that provide significant benefits in development. On the other hand, although developing countries are able to increase physical capital accumulation, they have not been able to catch up with developed countries. In this case, the theory of endogenous growth explains why capital accumulation does not have diminishing returns, but instead is increasing returns with specialization and investment in human resources (Meier and Rauch, 2000).

Indicator of Economic Growth

Economic growth is the concept of measuring changes in output that increase in the long term; growth is directly related to the increase in the production of goods and services measured by the increase in production output which ultimately increases national income through GDP. Economic growth can be measured by several indicators including (1) income growth (gross domestic product) (Suparmoko, 1990). According to Syahputra (2017), production capacity will increase if additional investment has an effect through aggregate demand and supply.

Growth will create prosperity through increased consumption due to higher income increases, therefore in measuring growth indicators of growth in production, income and operating profit are also used for the business sector. People's ability to consume is very dependent on their income; therefore income is a determinant of consumption (Mankiw, 2007), and competitive firms will maximize operating profits until the marginal product of labour equals real wages. (2) The growth of operating profit is based on the concept of Maupa (2004), that aggressive and innovative business owners will try to maximize profits and increase production to ensure long-term business continuity, meaning profit growth and production growth are important benchmarks in determining business growth.

The availability of financial resources, including those obtained from business profits, is a barometer of business growth. Another study says that the company will be able to survive in the long term if it is supported by good profit growth; profit growth will be achieved if revenue growth is higher than cost growth through production growth; then revenue growth and profit growth are indicators of business growth (Foreman-Peck et al., 2006). This study uses indicators of operating profit growth and revenue growth as a proxy for SHI growth.

Technology Usage

Technology etymologically comes from the Greek word "techne" which means the activity and expertise of carpentry by hand (Heidegger, 1977). Technique is a way of revealing or the way in which things are helped to appear. To express, what is the essence of technology? The response is crucial. The distinction between current technology and ancient/traditional technology is that modern technology does not entail anything in the poetic sense of a result beyond itself, such as production. While old technology contains creative qualities, such as the ability to create poetry. The challenge/challenge-forward disclosure is the most common in current technologies (Heidegger, 1977).

This way of disclosure requires nature to donate its energy excessively so that humans can store and use it. An example is a windmill which is a traditional technology that does not challenge the wind; windmills only rotate when there is a gust of wind, so the rotation is very dependent on the wind, so windmills only reveal wind energy but do not control energy for future use. While mining is challenging to the earth, the earth is exposed as a supply in the form of mines that produce energy (Lim, 2008).

All types of technology are embedded in culture because technology is created by humans (Ihde, 1979). Technology is a certain way of looking at the world (Heidegger, 1977). Technology was created as a mediator between humans and nature; technology can also be considered as a way of thinking that manipulates and exploits the world as a supply. The use of technology is also strongly influenced by the culture of the people; the same technology will be used in different ways depending on the culture of the people who use it (Lim, 2008). For example, gunpowder was first discovered in China and used as entertainment in fireworks events, but in western countries gunpowder is used as a weapon to conquer other countries. Because the technology created also contains elements of the culture of its creator, the transfer of technology from one country to another is directly followed by the transfer of culture and values belonging to the country of origin, because the receiving country technology has not been able to create new technology. So slowly but surely the transfer of technology will also be accompanied by changes in culture and values. So the technology is not free from value, but tends to direct value (Ihde, 1979). Humans are born with various physical limitations, so with the help of technology human abilities can be increased, such as the possibility to move quickly with modern means of transportation, or hear at a distance. This means that humans can transcend physical limitations through their imagination and creativity with the creation of technology.

The use of more advanced technology has implications for business efficiency, which can increase output with a constant amount of input use or decrease the use of inputs to produce a fixed output. In production activities, companies with business strategies based on technological capabilities will be able to compete in a competitive business world, as well as increasing the added value of production through increasing business scale by increasing the use of technology. Improving technological capabilities is related to increasing the contribution of the components that make it up, which includes the physical components of the equipment used, human skills, information technology, and the role of the organization. The contribution of each technology component in each activity forms the total technology contribution which is the level of technology. This level of technology payload is an indication of the strengths and weaknesses associated with technology and the dynamics of transformation. The four components must support each other, so that they are able to transform resources into production outputs that have higher added value more efficiently, which will indirectly improve business performance.

the results of Kuznet's research in 1966 (Amir, 2008) in the United States show that the contribution of technological progress to the economic growth of the United States in the 1929–1957 period was 78% and in the 1950–1962 period it was 56%. Amir (2008) shows the same trend, that the contribution of technological progress to economic growth in several developed countries is also not much different from the results of the Kuznet's study, around 49 to 76%. However, traditional technology, which is widely used in the SHI production process, does not have to be completely replaced with modern technology, because not all types of work can be done with modern technology. Technology is not only about intelligence and skills in using tools, but has expanded and includes all methods that are achieved rationally and have absolute efficiency for a certain stage of development in every field of human activity (Nala, 1989). Thus, the role of technology has changed behaviour, actions, and spontaneous human activities into actions and activities that have been calculated scientifically. The selection of technology must consider the type of technology that is most appropriate to solve problems and is highly efficient for the user community, and is able to optimally utilize natural and human resources. This study uses two indicators to reflect technology, namely traditional technology and modern technology.

Small Craft Industry

SHI is defined in various ways from various points of view by several experts and related agencies. The Ministry of Industry through the Decree of the Minister of Industry No. 286/M/SK/10/1989 and Bank Indonesia define small businesses based on the value of their assets; what is meant by small businesses are businesses whose assets (excluding land and buildings) are worth less than Rp. 600 million. Meanwhile, the Ministry of Trade limits small businesses based on their working capital; small businesses are businesses (trading) whose working capital is less than IDR 25 million.

The Indonesian Chamber of Commerce and Industry (Kadin) distinguishes small businesses into two groups, namely: (1) businesses engaged in trade, agriculture, and industry, (2) businesses engaged in construction. What is meant by small businesses for the first group are those with a working capital of less than Rp. 600 million. As for the second group, small businesses are those with a working capital of less than IDR 250 million and a business value of less than IDR 1 billion. From the point of view of net worth owned and annual sales value (State Ministry of Cooperatives and Small and Medium Enterprises), a small business is a business entity that has a net worth of at most Rp. 200,000,000, excluding land and buildings for business premises, and with annual sales of a maximum of Rp. 1,000,000,000.

Another definition is from the point of view of the number of workers used in the production process (Central Bureau of Statistics); a small business is a business entity that has a workforce of 5 to 19 people. The Ministry of Cooperatives and Small and Medium Enterprises differs (Law No. 20 of 2008), that a small business is an entity that has the following criteria.

1. Net assets of more than Rp. 50,000,000.00 (50 million rupiahs) up to a maximum of Rp. 500,000,000.00 (500 million rupiahs) excluding land and buildings for business premises.
2. Annual sales of more than Rp. 300,000,000.00 (300 million rupiah) up to a maximum of Rp. 2,500,000,000.00 (2 billion 500 million rupiah).

SHI is part of the people's economy (Mubyarto, 1997), namely as an economic system based on people's power which has the following characteristics: (1) done by the people without big capital; (2) managed by means of self-help; (3) independence; (4) no workers and no employers; and (5) not pursuing profit. The substance of the SHI lies on the constitutional basis of the people's economy, namely Article 33 of the 1945 Constitution, in the first paragraph of the explanation section which includes three elements (Baswir, 2009).

The three elements are as follows: (1) participation of all members of society in the national production process; (2) participation of all people in enjoying the results of national production. This is in line with Article 27 paragraph 2 of the 1945 Constitution and Article 34 of the 1945 Constitution; and (3) the activity of production and the sharing of its products must have ownership of community members. Empowerment of the SHI has a high moral message, which includes efforts to improve welfare through empowering skilled and unskilled workers, increasing production capacity at the rural level so as to reduce the rate of urbanization of the productive age population, increasing national production capacity, and being able to increase foreign exchange earnings through the export market. All of these activities form a dependency network from upstream to downstream that is able to encourage multiplier effects and increase national economic growth.

Based on the perspective of its development, SHI can be classified into four types according to business performance (Rahmana, 2009). The four classifications in question are as follows: (1) livelihood activities are businesses that are used as job opportunities to earn a living, which are more commonly known as the informal sector; (2) a micro enterprise is a business of craftsmen but does not yet have an entrepreneurial nature; (3) a small dynamic enterprise is a business that has an entrepreneurial spirit and is able to accept subcontracting and export work; and (4) a fast moving enterprise is a business that already has an entrepreneurial spirit and will transform into a big business (Amalya and Widyaningsih, 2018).

SHI also has specific characteristics and is very different from large business groups (Tambunan, 2006), as follows: (1) there are many businesses and they are scattered in rural areas; (2) it is labour intensive and closely related to the agricultural sector; (3) it uses relatively simple technology; (4) many of them can grow significantly so that they can develop into big businesses; (5) although rural residents are generally low-income, they are able to save some of their income to invest in business activities; (6) most of the business funds they manage come from their own capital; (7) the products produced are consumed by all groups of people, especially because the goods produced are simple goods; and (8) one of its advantages is its flexibility in production activities compared to large businesses.

There are at least four main reasons why SHI and other types of small businesses in Indonesia should be developed (Yustika, 2007). The four reasons are as follows: (1) the business structure in Indonesia is actually based on the existence of small/household/medium industries; (2) there are quite a lot of export-oriented industries, which really helps the country in earning foreign exchange; (3) the small business sector has proven to be more flexible in various unfavourable economic conditions; and (4) small business industrial products use more domestic raw materials or intermediate materials. SHI in this study uses Central Bureau of Statistics criteria, namely businesses that use between 5 to 19 workers.

Methodology

This is quantitative research which is based on the philosophy of positivism; it is aimed at testing general theories to produce specific findings through hypothesis testing, collecting data, and data analysis (Sugiyono, 2010), which ends with generalizations. The research objects are: (1) business growth with indicators: Revenue growth and profit growth. (2) technology consisting of traditional and modern technology, and traditional technology components consisting of indicators like carving equipment, basic pattern-making equipment, and paint equipment. The components of modern technology consist of indicators like machines, information communication equipment, and transportation equipment. This study used a sample of 95 SHI; the method of determining the number of samples was stratified proportionate random sampling, while the method of selecting respondents was simple random sampling (Rengiah and Sentosa, 2016). The data from the field research were obtained through a questionnaire instrument, while the data analysis used structural equation modelling (SEM) based on partial least squares (PLS).

Findings and Discussions

The average age of the respondents was 42.18 years; the youngest was 23 years and the oldest was 75 years. Age classification shows that the most dominant age is 40 years, meaning that the age of the respondent is closely related to work productivity. SHI is a modest industry that takes age and needs great morale, energy, and creativity. Of the total respondents that were analysed, it appears that almost 15% are under the age of 12 based on the average number of years in business for SHI, which is 14.80, with the youngest responder being 1 year old and the oldest being 49. This activity is the main source of income. In terms of education, the dominant business owner is an entrepreneur who graduated from high school (SLTA); the lowest level of education is elementary school and the highest is undergraduate (S1). The existence of SHI does not require a high level of formal education; high school graduates can very adequately manage a business; the most important thing for SHI is that entrepreneurs have a good basis, skills, experience, and understanding of the type of business being managed. Skills in the craft sector are much more important than high formal education. The average number of workers used by SHI is 7.21 people; the lowest number of workers is 5 people and the highest is 19 people. The average sales per month are Rp. 8.21 million. The lowest sales are Rp. 2 million, and the highest Rp. 20 million per month. Furthermore, based on the results of the calculation of the field data, the results are as shown in Table 10.1.

The calculation results in Table 10.1 show that the dimensions of modern technology and traditional technology are able to reflect technology constructs significantly. The dimension of traditional technology has an effect of 0.886 and is significant on the technological construct, while the dimension of modern technology has a smaller effect on the technological construct, which is 0.767 and is significant. This means that the role of traditional technology is more dominant than modern technology. The results of the next analysis show that technology has a positive effect of 0.404 on business growth and is significant at the 0.05 level. The results of this analysis can be seen further graphically in Figures 10.1 and 10.2.

The results of the next analysis show that technology has a positive effect of 0.11 on business growth but is not significant at the 0.05 level. This is because the technology required is generally relatively simple, which is more dominant than traditional technology. They receive raw materials in a form that is ready to be worked on, so they don't do a lot of engineering of raw materials with modern technology. The second reason is that works of art cannot be done with modern technology such as machines; works of art take advantage of taste which is an expression of one's artistic soul which is different from others, and more must be done with a touch of the hand through the help of traditional technology. Besides, the artistic spirit has a unique character, which is difficult to imitate and cannot be forced so that it is relatively difficult to mass produce, because it is closely related to the satisfaction of the workers who work on it.

The results of this study are different from western theories and the results of previous studies regarding the importance of the role of technological improvement, including neoclassical theory (Todaro and Smith, 2006), Kuznets' 1966 study in the United States (Amir, 2008), Solow (Baier et al., 2006) in

Table 10.1 Path analysis and T-test

Construct/indicator	*Original sample (O)*	*T Statistics (|O/STER)*	*Remark*
Technology → Growth	0.404	4.788	***Significant***
Technology → Modern	0.767	12.015	***Significant***
Technology → Traditional	0.886	35.950	***Significant***

(Source: The results of the calculation of the 2020 field survey data with PLS.)

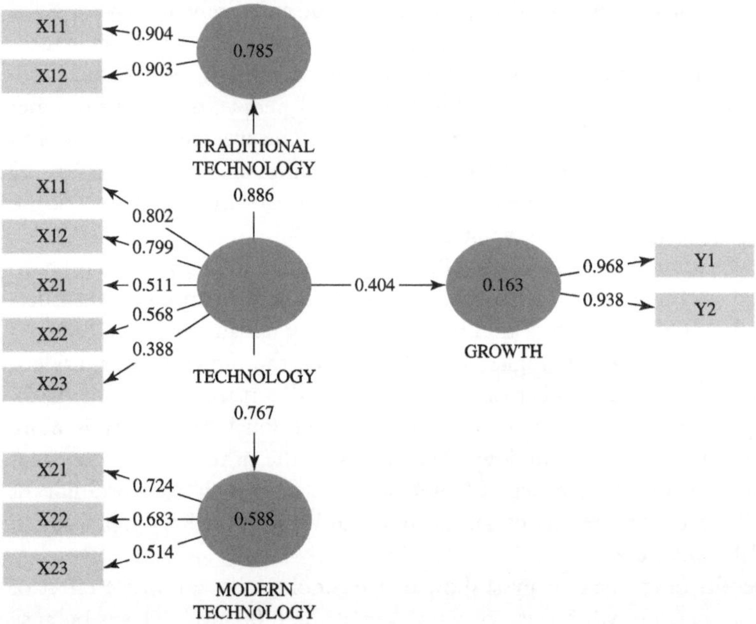

Figure 10.1 Technology usage and business growth model (source: The results of the calculation of the 2020 field survey data with PLS).

America between 1900 and 1949, Levy's study as in Husni (1994) in Iraq between 1961 and 1967, (Restuccia, 2004), Field in South Korea in 1961–1980 (Husni, 1994), and Amir (2008) in five developed countries, namely France, Japan, West Germany, England and the United States (Nakamura, 1981; Thatcher and Oliver, 2001; Kerr and Newell, 2003; and Wijono, 2005).

This study supports previous studies conducted in Indonesia by Amir (2008), and Bosworth, Collins, and Chen (1995) using data from 1964 to 1980. According to Amir (2008) technology is not the main variable in increasing business growth, but human resources are more important than technology. These results reinforce the position that the role of labour production factors is more dominant in encouraging business growth than technology. Capital provides a higher effect through labour production factors, meaning that the SHI is more labour intensive. This is very much in line with the reality on the ground, that SHI is a skill-based industry based on an artistic spirit, where indicators of talent and family environment are very important.

Conclusion and Recommendations

Based on the results of this analysis, it can be concluded that technology has a positive effect of 0.11 on SHI business growth in Bali Province but is not significant at the 0.05 level. This is because the role of the expertise of the workforce that comes from natural talent and the family environment is more dominant in

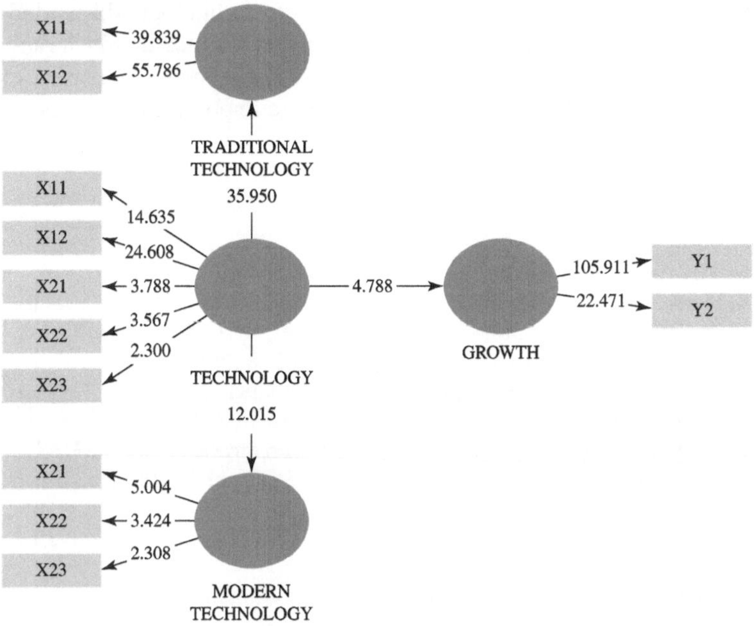

Figure 10.2 Bootstrapping model of business growth (source: the results of the calculation of the 2020 field survey data with PLS).

Table 10.2 Outer loading results

	Original Sample (O)	*Sample Mean (M)*	*Standard Deviation (STDEV)*	*T Statistics (|0/ STDEV|)*	*PP Values*
X11 <-TRADITIONAL TECHNOLOGY	0.904	0.902	0.023	39.839	0.000
X11<-TECHNOLOGY	0.802	0.801	0.055	14.635	0.000
X12 <- TRADITIONAL TECHNOLOGY	0.903	0.903	0.016	55.786	0.000
X12<-TECHNOLOGY	0.799	0.800	0.032	24.608	0.000
X21 <- MODERN TECHNOLOGY	0.724	0.698	0.145	5.004	0.000
X21<-TECHNOLOGY	0.511	0.497	0.135	3.788	0.000
X22 <- MODERN TECHNOLOGY	0.683	0.648	0.199	3.424	0.001
X22<-TECHNOLOGY	0.568	0.545	0.159	3.567	0.000
X23<-MODERN TECHNOLOGY	0.514	0.512	0.223	2.308	0.021
XZ3<-TECHNOLOGY	0.388	0.390	0.169	2.300	0.022
Y1 <- GROWTH	0.968	0.969	0.009	105.911	0.000
Y2 <- GROWTH	0.938	0.932	0.042	22.471	0.000

(Source: The results of the calculation of the 2020 field survey data with PLS.)

supporting business growth. The use of technology turns out to be more dominant with traditional technology compared to modern technology; this is due to the fact that SHI in Bali is more dominant in using a touch of art which cannot be helped much by modern technology that uses engine power or batteries. The element of skill with a touch of the hand is more important than modern technology.

References

Allen, R. G. D. (1967). *Macro-Economic Theory*. London: Macmillan and Co. Ltd.

Amalya, M. D. and Widyaningsih, T. W. (2018). Implementation of Naive Bayes for classification and potentially MSMEs analysis. *MATEC Web of Conferences*, 218, pp. 1–8. https://doi.org/10.1051/matecconf/201821802006

Amir, A. (2008). *Sumber-Sumber Pertumbuhan Ekonomi Indonesia*. Laporan Penelitian: FE/UNJA.

Asia Pacific Foundation of Canada (2018). *Survey of Entrepreneurs and MSMES in Indonesia*. Retrieved from: http://www.asiapacific.ca (accessed: 22nd December, 2021).

ASIAN Productivity Organization (APO) (2019). *ASIAN Productivity Organization Databook*. Tokyo: APO.

Baier, S. L., Dwyer, G. P. and Tamura, R. (2006). How important are capital and total factor productivity for economic growth. *Journal of Economic Inquiry*, 44(1), pp. 23–49.

Bank Indonesia (2010). *Regional Economic Study of Bali Province, Quarter 2 of 2010*. Bali: Bank Indonesia.

Barro, R. J. and Sala-i-Martin, X. (1999). *Economic Growth*. New York: MIT Press.

Baswir, R. (2009). Industri Kecil dan Konglomerasi di Indonesia: Prospek Kemitraan. *Prisma*, 10 Tahun XXIV. Oktober.

Bosworth, B. P., Collins, S. M. and Chen, Y. (1995). *Accounting for Differences in Economic Growth*. Brookings. Retrieved from: https://www.brookings.edu/research/accounting-for-difference-in-economic-growth/ (accessed: 2nd December, 2021).

Central Bureau of Statistics (2010). *Bali in Figures*. Bali: Central Bureau of Statistics.

Central Bureau of Statistics (2016). *Economic Census of Bali Province*. Bali: Central Bureau of Statistics.

Central Bureau of Statistics (2018). *Bali in Figures*. Bali: Central Bureau of Statistics.

Central Bureau of Statistics (2019). *Bali in Figures*. Bali: Central Bureau of Statistics.

Director General of Indonesian Small and Medium Industries (2018). *Small and Medium Industry Echo. Edition 60 March, 2018*. Jakarta: Director General of Indonesian Small and Medium Industries.

Duan, Y., Mullins, R., Hamblin, D., Stanek, S., Sroka, H., Machado, V. and Araujo, J. (2002). Addressing ICTs skill challenges in SMEs: Insights from three country investigations. *Journal of European Industrial Training*, 26(9), pp. 430–441.

Foreman-Peck, J., MakePeace, G. and Morgan, B. (2006). Growth and profitability of small and medium sized enterprises: Some Welsh evidence. *Regional Studies*, 40(4), pp. 307–319.

Heidegger, M. (1977). *The Question Concerning Technology and Other Essays*. New York and London: Garland Publisher Inc.

Husni, E. (1994). *Fungsi Produksi Cobb-Douglas Pada Industri Manufaktur di Indonesia*. Thesis. Jawa Barat: Universitas Indonesia.

Ihde, D. (1979). *Technics and Praxis: A Philosophy of Technology*. Dordrecht: Reidel Publisher.

Kerr, S. and Newell, R. G. (2003). Policy induced technology adoption: Evident from the US lead phasedown. *The Journal of Industrial Economic*, 1, pp. 1–28.

Ladokun, I. O. and Osunwole, O. (2013). Information and communication technology in small and medium enterprises: Factors affecting the adoption and use of ICT in Nigeria. *International Journal of Academic Research in Economics and Management Sciences*, 2(6).

Layard, P. R. G. and Walters, A. A. (1978). *Microeconomic Theory*. New York: Mc Graw-Hill Book Co.

Lim, F. (2008). *Filsafat Teknologi*. Yogyakarta: Kanisius.

Mankiw, N. G. (2007). *Macroeconomics*. New York: Worth Publishers.

Maupa, H. (2004). Faktor-faktor Penentu Pertumbuhan Usaha Kecil di Sulawesi Selatan. *Analisis*, 1(2).

Meier, G. M. and Rauch, J. E. (2000). *Leading Issue in Economic Development*. Oxford: Oxford University Press.

Mubyarto. (1997). *Ekonomi Rakyat, Program IDT, dan Demokrasi Ekonomi Indonesia*. Yogyakarta: Aditya Media.

Nakamura, T. (1981). *The Japanese Economy its Development and Structure*. Tokyo: University of Tokyo Press.

Nala, N. G. (1989). *Penerapan Teknologi Tepat Guna di Pedesaan*. Bali: P3M Unud.

Prabawa, A. A. N. P. and Budhi, I. M. K. S. (2017). Pengaruh Modal, Tingkat Upah, dan Teknologi Terhadap Penyerapan Tenaga Kerja dan Produktivitas Pada Industri Sablon di Kota Denpasar. *E-Jurnal Ekonomi Pembangunan UNUD*, 6(7), pp. 1157–1184.

Rahmana, A. (2009). Peranan Teknologi Informasi dalam Peningkatan Daya Saing Usaha Kecil Menengah. In Seminar Nasional Aplikasi Teknologi Informasi (SNATI) Yogyakarta, 20 Juni 2009.

Rengiah, P. and Sentosa, I. (2016). The effectiveness of entrepreneurship education in developing entrepreneurial intention among Malaysian University Students: (A research findings on the structural equation modeling). *European Journal of Business and Social Sciences*, 5(2), pp. 30–43.

Restuccia, D. (2004). Barrier to capital accumulation and aggregate total factor productivity. *Journal International Economic Review*, 45(1), pp. 238–255.

Romer, P. (1986). Increasing return and long growth. *Journal of Political Economy*, 94(5), pp. 1002–1037.

Simangunsong, F., Hutasoit, I. and Sentosa, I. (2019). A strategic framework of good governance, infrastructure development and community empowerment in Indonesian public sector management. *African Journal of Hospitality, Tourism and Leisure*, 2019(Special Issue), pp. 1–12.

Sugiyono. (2010). *Metode Penelitian Kuantitatif Kualitatif & RND*. Bandung: Alfabeta.

Suparmoko, I. M. (1990). *Ekonomi Pembangunan*. Penerbit: Yogyakarta-BPFE.

Syahputra, R. (2017). Analisis Faktor-Faktor Yang Mempengaruhi Pertumbuhan Ekonomi Di Indonesia. *JURNAL SAMUDRA EKONOMIKA*, 1(2), pp. 183–191.

Tambunan, T. (2006). Analisis Keterkaitan Klaster. *Jurnal Pengkajian Koperasi dan UKM*, 2(Tahun I).

Thatcher, M. E. and Oliver, J. R. (2001). The impact of technology investment on a firm, production efficiency, product quality and productivity. *Journal of Management Information System*, 18(2), pp. 17–45.

Todaro, M. P. and Smith, S. C. (2006). *Pembangunan Ekonomi Jilid 1*. Jakarta: Jakarta c Erlangga.

Wijono, W. (2005). Mengungkap Sumber-Sumber Pertumbuhan Ekonomi Indonesia Dalam Lima Tahun Terakhir. *Jurnal Manajemen dan Fiskal*, V(2), pp. 1–17.

Yustika, E. (2007). *Perekonomian Indonesia: Satu Dekade Pasca Krisis Ekonomi*. Malang: BPFE/UNIBRAW.

11 Technological Policy Guidance on Rural Tourism Sustainability in Indonesia

Tri Wahyuningsih, Ilham Sentosa, and Sheikh Muhamad Hizam

Introduction

The awareness that the environment has intrinsic value that exceeds its value as a tourism asset is the cornerstone of sustainable development. The growth of tourism can accelerate the destruction of natural resources. Sustainability development is threatened by the inability to preserve cultural heritage (Bhati and Pearce, 2017). This is especially concerning for destinations that rely on ecological characteristics that make them vulnerable to becoming a tourist attraction (Giglio et al., 2018), or are located in areas where infrastructure, legislation, and existing experience are insufficient to handle the impact of increased visitor numbers, and these negative consequences may have an impact on the environment's long-term viability (Hashemi and Ghaffary, 2017). Developed countries now have made integrated steps to protect and protect the ecology and nature (He et al., 2018). This is supported by the recognition of tourists of their impact on the destinations they visit (Weaver and Jin, 2016).

The four principles of sustainable tourism development are economic feasibility, environmental feasibility, social acceptableness, and technological feasibility (Loi et al., 2014). The economically feasible principle, which highlights the importance of financial viability in the development process, is efficiently executed in order to provide actual economic benefits for regional growth and local community welfare. The notion of ecologically feasible development emphasizes that the development process must be sensitive to attempts to maintain the environment (natural and cultural) and, at the very least, minimize negative impacts that can lower environmental quality and disrupt the ecological balance (Loi et al., 2014). The idea of socially acceptable development places a strong emphasis on the fact that development activities must be carried out while taking into account the values and norms that exist within the community and that development impacts should not disturb them. The idea of technologically suitable development emphasises the need of choosing a development plan that is technically possible, effective, and easily accepted by local populations for a long-term management approach because it makes use of local resources (Osman and Sentosa, 2013).

In simple terms, sustainable tourism development can be broken down into three main goals: (1) Quality of natural and cultural resources, where tourism

DOI: 10.4324/9781003315179-15

development must protect the integrity of existing natural and cultural resources while also taking into account the region's capacity to accept/tolerate tourism development. (2) The socio-economic quality of life of the local community, where tourism development must have a beneficial influence (benefit) on socio-economic status of the local community, such as increasing job possibilities or even making it economically self-sufficient. (3) Tourist experience quality (tourists), where tourism development must be responsive to the level of tourist satisfaction, thereby making the tour a valuable experience. The quality of tourist items and their interpretation play a critical impact in the quality of one's trip experience in this scenario (Jader and Santosa, 2015).

The Tri Daya concept, which entails the development of a tourist village, is well-known in Indonesia. Tri Daya is a notion of human empowerment that incorporates social, environmental, and economic power. The purpose of social power is to improve a community's social power so that a more effective society may be established. Economic power tries to generate empowerment in order to form a productive society, whereas environmental power aims to generate development power in order to create a sustainable environment (Purnomo et al., 2020).

Rural Tourism in Indonesia

Indonesia has 74,954 villages, 1,902 of which are being developed as tourism villages, indicating that tourism villages in Indonesia will face more competition in the future. There is fierce competition in the tourism industry, notably in Central Java, where there were 147 tourist villages in 2016; of them, 75 were classified as mainstay tourism, 50 as superior, and 22 as deserving of being turned into tourism villages (Purnomo et al., 2020). Tourism villages in Yogyakarta are also a rival to the tourist villages in Central Java. This is a challenge for tourism managers in the village to win the competition. This is one of the community-based tourism development strategies. One representation of community-based tourism development is the tourism village. The concept of tourism development is still a hot topic of discussion in the media, as well as the themes of research and community service both among nongovernmental organizations and universities.

Recent tourism research has focused on multinational hotel development (Maclean et al., 2018), historical and ancestral tourism (Murdy et al., 2018), and ecotourism development (Murdy et al., 2018). Nonetheless, historical research in the hospitality and tourism industries is still in its infancy (Heslinga et al., 2018). Current academic attention has focused on investigating issues around the objectives, services, and sustainability of the site from a multi-stakeholder perspective; this is usually centred on exploring the manifestations of current sustainable practices (Boley et al., 2017; Jones et al., 2016). Therefore, to complement the focus in this field, research investigates the factors that underlie sustainable tourism to get a more holistic picture around sustainable development in general (Foster et al., 2017).

Based on the description above, a lot of research on sustainable tourism has been done, but research on rural tourism is still rarely found, especially about the impact of managing sustainable tourist destinations on tourist loyalty. Asmelash and Kumar (2019) examined the relationship of the dimensions of sustainability based on economic, environmental, socio-cultural and institutional factors and tourist satisfaction with museum/cultural tourism; the results of their research indicate that economic, socio-cultural, and institutional sustainability have a significant effect on tourist satisfaction, whereas environmental sustainability does not. Then, a study undertaken by Agyeiwaah, McKercher, and Suntikul (2017) discovered seven indicators: job creation, firm viability, quality of life, water quality, waste management, energy conservation, and community integrity maintenance.

Research of Campón-Cerro, Hernández-Mogollón, and Alves (2017) examined the competitiveness of sustainable tourism villages; the results of their research found that image, quality, and satisfaction significantly influence tourist loyalty. Image, quality, and value have a significant effect on satisfaction. These results support research conducted by Cong (2016), and Al-Ansi and Han (2019) who examine the effect of perceived quality and satisfaction on tourist loyalty as measured using word of mouth and revisit intention. The findings of Cong (2016) reveal that tourists' perceptions of the quality of tourist locations have a considerable impact on contentment, and satisfaction has a large impact on word of mouth and intent to return.

Ardyan and Wibisono (2019), Suhartanto, Brien, Primiana, Wibisono, and Triyuni (2020), and Sangpikul (2018) studied the impact of travel experiences on tourist satisfaction and loyalty. Their findings revealed that travel experiences had a substantial impact on tourist satisfaction and loyalty. This shows that a memorable travel experience will increase tourist satisfaction and tourist loyalty to visit again and recommend to others. Tourist satisfaction is a focus of research on tourists' impressions of locations and future behavioural intentions, as it influences their decision to return (Sangpikul, 2018; Suhartanto et al., 2020; Cong, 2016). Another essential component is word of mouth recommendations, which are made by tourists who visit or suggest the location to others, as opposed to unsatisfied travellers (Suhartanto et al., 2020; Cong, 2016).

Many of the results of these studies contradict each other, if satisfaction is used as a mediating variable. Several studies have found that satisfaction has a full, partial, or no mediation influence on the relationship between the quality of objectives and behavioural intentions or loyalty behaviours (Loi et al., 2014; Žabkar et al., 2010; Al-Ansi and Han, 2019). Prior experience is a socio-demographic feature that has been studied extensively in recent studies. Choi, Kim, and Cai (2012) and Rajaratnam, Nair, Sharif, and Munikrishnan (2015) revealed that previous visits reinforced the desire to return.

The main purpose of this research is to examine the relationship between tourist perceptions of the sustainability of rural tourism in Indonesia and tourist satisfaction and tourist behaviour. This research examines the influence of moderation from previous visits and the origin of tourists, namely local, regional, or international tourists. Because the COVID-19 pandemic has had an influence on many sectors, including the tourist industry, current research on sustainable tourism cannot disregard the impact of the ongoing COVID-19 pandemic and regulations after the COVID-19 pandemic. Previously, the tourism business in Indonesia had been growing rapidly as the hashtag "Wonderful Indonesia" became well-known around the world, but during the pandemic, both domestic and international travellers saw a dramatic drop in numbers. When compared to other industries, the tourist industry is one of the most severely hit by the pandemic. Indonesia has lost around US$1.5 billion in the tourism business since January (World Travel & Tourism Council, 2020).

The Minister of Tourism and Creative Economy, Republic of Indonesia, Dr. Sandiaga Uno said tourism was the sector most affected by the coronavirus pandemic or COVID-19. This is because the number of tourists visiting the country is not only reduced, but almost non-existent. In 2019 the number of tourists visiting Indonesia from abroad was 16 million visitors. They provided foreign exchange to the country of US$20 billion (Okezone.com, 2021). The most devastating effect of the 2019 coronavirus disease is on human health. However, it also has a major impact on the economy (Huang, 2020). Industry has been beset by uncertainty since the COVID-19 breakout in late 2019, and this scenario is particularly obvious in the tourism and hotel industries. The tourism and hospitality industry, which depend on people moving about and interacting closely, are both co-creators and main benefactors of the pandemic recovery strategies (Gallen, 2020).

According to research conducted by Hao, Xiao, and Chon (2020), the growth of business channels, product design and investment preferences, digital and smart transformation, and market overhaul are the primary themes in China's hotel industry post-pandemic. This paper shows that hotel organizations worldwide must comprehend the catastrophic scenario, develop an effective anti-pandemic strategy, and aim for industrial activation and revitalization, based on disaster management literature and experiences and trends in the Chinese hotel sector. Despite the fact that China's hotel business was the first to feel the effects of the pandemic, quick action was required to avoid economic losses and protect employees and customers. During a comprehensive and lengthy lockdown, the industry practices social responsibility and actively engages in the anti-pandemic campaign (Hao et al., 2020).

Many studies have been conducted since the COVID-19 pandemic, particularly on the impact of the pandemic on tourism, but few studies have specifically examined the impact of the pandemic on the continuation of tourist villages, whose management is still primarily carried out in a simple manner. Therefore, this study will examine the sustainability of tourist villages, especially in terms of the impact of sustainable tourist destination management on tourist loyalty in this pandemic era and the strategies undertaken by tourist villages to overcome the impact of the COVID-19 pandemic.

Challenges for the Rural Tourism in Indonesia

Research on rural tourism is still rarely found, specifically about the impact of sustainable tourism destination management on tourist loyalty. Economically feasible, environmentally feasible, socially acceptable and technologically appropriate are the four criteria that underpin the concept of sustainable tourism development. Village tourism managers must be able to integrate the four principles so that rural tourism can be sustainable or well developed. The application of sustainability principles has an impact on the quality of rural tourism management. Good service raises a sense of satisfaction for tourists on their visits to the tourist village. Satisfaction provides a pleasant experience for tourists, which leads to loyal behaviour that is shown in the desire to return and recommend to others.

The research aims to analyse the implementation of sustainable tourism village management. This research is important for the Indonesian government to develop rural tourism throughout Indonesia, so that tourism village managers can manage rural tourism in a sustainable manner. The economic, socio-cultural, environmental, and institutional components of the notion of sustainable tourist development must all be appropriately applied. For the tourism village, this research is expected to provide a reference in developing rural tourism so that the villagers will be more prosperous.

The study question asks whether there is a connection between the three constructs of sustainable tourism, tourism marketing, and satisfaction, with experience quality and tourists type serving as moderating factors. As a result, this research adds to the theoretical and practical elements of the causal link between these constructs. Theoretically, this study supports the empirical influence of independent variables on dependent variables, which supports the preceding. This study contributes to the development of existing research in Indonesia, especially on the tourism industry, by showing the effect of the tourism sustainability component on tourist satisfaction recommendations and return visits. Thus, this research will enrich the existing literature through sustainable tourism research, satisfaction, recommendations, and repeat visits.

The findings of this study will enable the Karanganyar Districts to develop rural tourism by applying the principles of sustainable tourism to all rural tourism. Recommendations to visit as a result of tourist satisfaction can help rural tourism marketing. The suggestions about weaknesses and strengths listed can help regulate changes in satisfaction, return visits, and marketing. For the rural tourism manager, this research is expected to provide input for sustainable rural tourism management, so that it can develop service innovations that can satisfy tourists visiting tourist attractions in the village, which in turn can increase tourist loyalty, so that the sustainability of tourism villages can improve the welfare of the community.

Rural Tourism Sustainability

Economic sustainability is a philosophy that emphasizes that the development of rural tourism must be financially possible and properly carried out in order to give substantial economic advantages both for regional development and for the enhancement of local populations' welfare. Dimensions of economic sustainability include employment quality, economic viability, and local prosperity (Asmelash and Kumar, 2019). Environmental sustainability is a philosophy that emphasizes that the development process must be sensitive to attempts to maintain the environment (both natural and cultural) and, at the very least, minimize negative impacts that can degrade environmental quality and disrupt ecological balance. Physical integrity, biological diversity, resource efficiency, and environmental purity are all aspects of environmental sustainability (Asmelash and Kumar, 2019).

The notion of socio-cultural sustainability emphasizes that the development process must be socially acceptable, that development activities must be carried out while paying regard to the community's values and norms, and that development impacts must not disrupt them. Socio-cultural sustainability dimensions include social equity, visitor fulfilment, local control, community wellbeing, and cultural richness (Asmelash and Kumar, 2019). Institutional sustainability is the principle of managing rural tourism by paying attention to institutional governance, community participation, and policies oriented to local communities in the long run. Local-oriented control policy, political engagement, local policy planning, and political support at various levels of government are all aspects of institutional sustainability (Asmelash and Kumar, 2019).

Tourists' psychological and social reactions to the performance of attractions determine the quality of their experience. This description demonstrates how the quality of service and the quality of experience differ conceptually. The excitement and fantasy of tourists with attractions are likely to trigger the quality of the service experience (Cetin and Bilgihan, 2016; Suhartanto et al., 2020). Because the quality of experience includes emotions and sensations that are not quantified in service quality, the scope of the customer experience is greater than the quality of service. Despite the fact that the quality of tourism experience has been extensively researched, various sectors of tourists can have varied experiences depending on their reasons to visit different places (Cetin and Bilgihan, 2016; Suhartanto et al., 2020). Tourist satisfaction is the perception of tourists visiting rural tourism areas. The indicators used include attractiveness, hospitability, tourist staff treatment, reasonability of entrance fees, quality of information, safety and security, accessibility, and accommodation (Asmelash and Kumar, 2019).

Based on the sustainability issue, it appears that there are issues with the management of sustainable rural tourism. As a result, studying the elements that influence the sustainability of rural tourism is highly intriguing. According to observations, the reason rural tourism is not as active or is less active is because tourists are not travelling to the rural tourism destination. In contrast, when rural tourism develops, tourist visits are increasing and some visitors go to the place more than two times. These initial findings need to be studied more comprehensively in relation to the management of sustainable rural tourism and their relationship with the level of satisfaction and loyalty of tourists which is a benchmark for the success of the sustainability of rural tourism.

Understanding Rural Tourism

Rural tourism, also known as ecotourism or agro-tourism, despite the fact that there is no scholarly or legal consensus on the notion, has been adopted by many countries throughout the world as one of the key rural strategies for creating money in the rural environment (Guzman-Parra et al., 2015). Rural tourism is an important political action outside of traditional rural policies dedicated to agriculture and livestock, according to studies for sustainable development in rural regions (Mataveli and Gil, 2018).

Rural tourism is the development of a village area that is more inclined to develop the potential of existing villages by incorporating the capabilities of the elements in the village that function as attributes of small-scale tourism products into a series of tourism activities or activities that are able to provide and fulfil a variety of tourist travel needs both in terms of attractiveness and as a supporting facility (Muljadi, 2012). Purnomo, Rahayu, Riani, Suminah, and Udin (2020) claim that distinctive and intriguing socioeconomic life or economic activities have the potential to provide a variety of tourism-related demands, such as attractions, lodging, food and drink, souvenirs, and other travel necessities. Three factors drive the development of rural tourism. First, rural locations have more true natural and cultural potential than urban ones. Rural communities continue to practice cultural activities, customs and rituals in a rather harmonious manner. Second, compared to metropolitan regions, rural areas have more natural physical environments or are less polluted by various sorts of pollution. Third, rural areas experience relatively slow economic development, making it difficult to fully utilize economic potential.

According to Putra (2006), rural tourism refers to a region or rural area that can be utilised based on the presence of a number of components that have integrated tourism product attributes. The rural area provides an overall atmosphere resembling a village with a theme of rural authenticity, both in terms of order and authenticity. A rural area that provides an overall atmosphere that reflects the authenticity of the countryside itself, starting with socio-culture, traditions, daily life, architectural structures, and village spatial structures, and culminating in a series of tourism activities and activities, is defined as rural tourism (Ismayanti, 2013).

In addition to this uniqueness, the area of rural tourism is also required to have various supporting facilities as a tourist destination. Some of these facilities will facilitate tourist village tourists in carrying out tourism activities. Some facilities that are usually in the area of the tourist village include the following: transportation, telecommunications, accommodation, and health facilities. For accommodation, the tourist village can provide a place to stay in the form of a home stay so that tourists can feel the authentic rural atmosphere (Ismayanti, 2013).

The economic benefits and long-term expansion of rural tourism have enormous potential to rebuild local economies, create jobs for family members and farmers and improve the quality of life in agrarian communities (Mataveli and Gil, 2018). Tourism has been considered as a potential source of rural development. There are various strategies to improve the wellbeing of people living in rural areas. In more detail, the impact of tourism in rural regions is evaluated in three ways. The initial examination is focused on the tourism's immediate consequences. The second effect is on the local economy in an indirect manner. The third estimate also shows a dynamic long-term impact on infrastructure and local environmental management (Mataveli and Gil, 2018).

Innovative systematisation for critical welfare-related factors is provided by Romo, Machino, and Nijkamp (2018). The expansion of tourism in rural regions, assuming that this sort of tourism is a brand-new niche that has to be studied as a location where other types of tourists are already present. This strategy is based on the concept of integrative diversification, which helps to clarify the interdependencies and connections between the many activities and players involved in regional tourist dynamics (Osman and Sentosa, 2013). This systematization aims to achieve three main goals: creating different and unique health-related services, enabling destinations to achieve strong competitive positions in the global market, and the ongoing promotion of the use of territorial resources for the quality of life of the local population, as well as for the preservation of competitive positions; achieving many benefits for local people, by mobilizing local resources to enhance entrepreneurship for tourism development; and achieving many benefits for local people, by mobilizing local resources to enhance entrepreneurship for tourism development. Based on some of the opinions of the experts above, it can be explained that the tourism village is an area that is a tourist attraction with characteristics such as natural beauty, cultural arts, and daily habits of the community where tourists can directly feel the life of the people in the village.

Rural Tourism and Technology Policy for Sustainability

Each tourism village certainly has its own characteristics to be used for tourism. The management of a tourist village as a tourist attraction is not only limited to its determination as a tourist village. The determination of a village as a tourist village is based on at least a number of potential supporting components. There are attractions that are typical of the village itself (Osman and Sentosa, 2013):

1. The existence of tourism facilities and accommodation such as lodging facilities, food and drink facilities, hawker centres or souvenirs, visitor centres;
2. The existence of tourist activities such as weaving, enjoying the scenery, and others;
3. The existence of general development as an effort to create a tourist destination that provides the best service for tourists, including division of zones or areas, management of visitors, and communication services.

To be designated as a tourist village, a village must meet a number of criteria, including (Sastrayuda, 2010):

1. Good accessibility, using numerous modes of transportation to make it easier for tourists to visit;

2. To be developed as a tourist destination, it must have interesting objects that may be natural, cultural arts, legends, local delicacies, and so on;
3. The community and local leaders provide unwavering support for the tourism village and visitors;
4. The village's safety is secured;
5. There is sufficient accommodation, telecommunications, and human resources;
6. There is a cool or cold climate;
7. They have a relationship with other attractions that are already known by the wider community.

The following are some of the factors that have an impact on the growth of a tourist village:

1. Human resource development (HR): HR can be developed through educational activities, training and participation in seminars, discussions etc.;
2. Partnership: good and mutually beneficial cooperation between the management of the tourism village and tourism entrepreneurs in the city or the relevant regional tourism office in several business fields, namely the fields of accommodation, travel, promotion, training, and others;
3. Government activities in the village: there are activities carried out by the village government, for example, official meetings, construction exhibitions, and traditional ceremonies held at tourist villages;
4. Promotion: rural tourism must be promoted through various media. Therefore, the village or the regency of the area around the tourist village must often invite reporters from print and electronic media to carry out promotional activities;
5. Festival/matches: periodically in the tourist village activities should be carried out that can attract tourists to visit the tourist village, for example holding art festivals or sports matches;
6. To provide guidance to community organizations: many villagers migrate elsewhere. However, they can also be organized and fostered to advance their tourist villages through community organizations and to reduce unemployment in the village.

According to Sastrayuda (2010), a tourism village must have a unique, original, characteristic location near extraordinary natural areas, be related to cultured groups or communities that primarily attract visitors, and have the opportunity to develop both in terms of basic infrastructure and other facilities. According to Putra (2006), the village must have tourism, arts, and culture potential that is typical of the area, the site of the village must be included in the scope of tourism development, or at the very least be included in the corridors and routes of tour packages that have been marketed, and management personnel, trainers, and performers in tourism, arts, and culture must be available (Osman and Sentosa, 2013). Infrastructure and accessibility enable the tourism village programme. Security, order, and cleanliness are all guaranteed.

According to the Brunlad Report of the World Commission on Environment and Development (WCED) in 1987, "sustainable development is growth that meets current demands without endangering future generations' ability to meet their own needs" (WCED, 2016: 2). From this remark, it is clear that sustainable development is an aspect of long-term development that considers current demands while also considering future generations' ability to meet those requirements. Similarly, the World Trade Organization (WTO) promotes development concepts such as ecological, social, and cultural sustainability, as well as economic sustainability, for both present and future generations (Anom, 2010).

The tourist industry has been growing since the mid-twentieth century, and it has continued to grow fast on a global basis without showing any indications of slowing down (Estêvão et al., 2019). This practice has put undue strain on tourist locations and their resources, potentially causing damage to their tourist attractions over time (Carayannis et al., 2018). Tourist destination sustainability policies have evolved into strategic planning for tourism entity operations in response to this problem, and they are becoming increasingly critical (Estêvão et al., 2019).

Over time, the concept of sustainable development was incorporated into the concept of sustainable tourism development. A tourism-oriented development method that ensures the long-term availability of resources essential for future development is known as sustainable tourism development. Tourism forms that are environmentally, socially, and communally conscious, and that allow both hosts and tourists to have a pleasant and beneficial interaction and shared experience are also included (Estêvão et al., 2019).

Sustainable business integration must consider economic, social, and environmental aspects (Darbari et al., 2019; Shankar et al., 2017; Li et al., 2019). In current economic activities, the concept of sustainable development is a hot topic (Dobrovolskienė et al., 2019; Govindan et al., 2018). However, by definition, the concept of sustainability is not universally accepted (Fernandes et al., 2018; Govindan et al., 2018). The basic principle in sustainable development, as a long-term process to ensure the demands of current and future generations can be satisfied, is the triple bottom line (economic equality, environmental preservation, and social justice) (Estêvão et al., 2019).

According to Kernel (2005), evaluating sustainable tourism is a difficult task due to its multi-dimensional character and environment, as well as the involvement of several stakeholders, many of whom have competing interests, making the evaluation of sustainable tourism a complicated problem. Sustainable tourism has far-reaching implications for a wide range of stakeholders, including the government and society. When fiscal constraints harm the region, these stakeholders' opinions frequently clash. In this setting, a complete integrative approach to assess sustainable tourism is required (Estêvão et al., 2019).

According to Budeanu, Miller, Moscardo, and Ooi (2016), in order to overcome some restrictions and contribute to advancement in this sector, researchers must investigate new approaches in the context of sustainable tourism. Current approaches for evaluating sustainable tourist regions, in particular, still have limitations in terms of how this assessment model chooses criteria to be included in the evaluation mechanism (Carayannis et al., 2018). Another issue is calculating trade-offs between criteria decisions, which is still a mystery due to a lack of understanding of the causal relationship between criteria and sustainable tourism features (Franzoni, 2015; Pérez Gálvez et al., 2017; Reis et al., 2019).

Furthermore, the Sustainable Tourism Charter underlines that tourism must be founded on sustainable principles, which means that tourism must be commercially viable as well as ethically and socially fair to local populations. Furthermore, according to Fitra and Leksmono (2001), the idea of sustainable development involves three interconnected components: first, the term "ecological sustainability" refers to the fact that tourism development should not cause irreversible changes to an existing ecosystem, and it has become a widely accepted component since natural resources must be preserved from the negative consequences of tourism. Second, social adaptability, which refers to a group's ability to absorb tourists without disrupting social ties, either between members of the community group and tourists or among group members. Third, cultural sustainability argues that the presence of tourists in a tourist site should not impede the growth of local culture, but that the culture's survival must be preserved for future generations (Jader and Santosa, 2015).

Two approaches to tourism are required to fulfil the goal of sustainable tourism development. Fitra and Leksmono (2001), showed two related models, including:

First, horizontal linking (horizontal circle); this method assumes that tourism will facilitate the implementation of numerous programmes and policies. Various supporting policy components are required for the process to be efficient, including the ability to clearly identify the problem, define the development vision and mission, comprehend the hierarchy of programme goals and objectives, and appropriately arrange the process. In this perspective, tourism is a part of a larger process that involves collaboration with other sectors. Second, vertical connecting is present (vertical circle). The purpose of integrating these strategies is to achieve a balance when combining critical components of tourism and development activities, as well as to "protect" creative decision-making breakthroughs. The characteristics of vertical relationships are as follows: first, in this approach, tourism is a part of development that functions as a strategic part of policymaking, so that it is at the top and influences other sectors; second, the strategic elements of policy planning must include the provision of tourism facilities and infrastructure; third, the growth of specialised tourism, which includes lodging of all kinds, including hotels, motels, and other; fourth, the estimated impact (including carrying capacity studies) of tourism development in terms of economic, environmental, local socio-economic, and cultural heritage; fifth, financing, marketing, promotion, and information systems; and seventh, a tourism awareness campaign for the community (Fitra and Leksmono, 2001).

Thus, the concepts proposed by Burns and Holder became a reference in the development of sustainable tourism, based on diverse perspectives and conceptual research (community-based tourism). In other words, sustainable tourism is a tourism concept developed by those who recognize the significance of the term "sustainability," emphasizing the long-term viability of a tourism area's development in three areas: environment, social culture, and economy. Therefore, excellent and integrated strategic planning is needed by all stakeholders in its implementation.

The concept of community-based tourism (CBT) was developed as a framework for enabling long-term tourism growth in rural areas (Hall and Page, 2014). CBT concentrates its efforts on the earliest stages of the process with local tourism promoters. It is specifically focused on increasing local awareness of the tourism business and inspiring local engagement in helping to improve tourism (Chatkaewnapanon and Kelly, 2019). When attempting to obtain CBT, it is best to begin by focusing on the behaviours and reactions of tourist initiators from local people who are seeing tourism development. This will enable them to create appropriate tourist activities since they will be well informed of tourism resources and their attraction (Chatkaewnapanon and Kelly, 2019).

Conclusion

The concept of sustainable tourism development refers to the application of the concept of sustainable technological development in the tourism sector, which in essence implies a tourism development understanding that is responsive to tourist interests and direct community involvement while emphasizing long-term protection and management efforts. Efforts to generate and manage resources must consider economic, social, and aesthetic aspects. It is possible to sustain the integrity and/or preservation of ecology, biodiversity, culture, and living systems at the same time. The present study conceptualized technological policy guidance on rural tourism sustainability in Indonesia.

References

Agyeiwaah, E., McKercher, B. and Suntikul, W. (2017). Identifying core indicators of sustainable tourism: A path forward? *Tourism Management Perspectives*, 24, pp. 26–33.

Al-Ansi, A. and Han, H. (2019). Role of halal-friendly destination performances, value, satisfaction and trust in generating destination image and loyalty. *Journal of Destination Marketing and Management*, 13, pp. 51–60.

Anom, I. P. (2010). *Pariwisata Berkelanjutan Dalam Pusaran Krisis Global.* Denpasar: Udayana University Press.

Ardyan, E. and Wibisono, U. (2019). Between self congruity, destination relationship and memorable tourist experience: An empiric study on the loyalty of tourist destination. *Jurnal Dinamika Manajemen*, 10(1), pp. 111–123.

Asmelash, A. G. and Kumar, S. (2019). The structural relationship between tourist satisfaction and sustainable heritage tourism development in Tigrai, Ethiopia. *Heliyon*, 5(3), p. e01335.

Bhati, A. and Pearce, P. (2017). Tourist attractions in Bangkok and Singapore; linking vandalism and setting characteristics. *Tourism Management*, 63, pp. 15–30.

Boley, B. B., McGehee, N. G. and Tom Hammett, A. L. (2017). Importance-performance analysis (IPA) of sustainable tourism initiatives: The resident perspective. *Tourism Management*, 58, pp. 66–77.

Budeanu, A., Miller, G., Moscardo, G. and Ooi, C. S. (2016). Sustainable tourism, progress, challenges and opportunities: An introduction. *Journal of Cleaner Production*, 111, pp. 285–294.

Campón-Cerro, A. M., Hernández-Mogollón, J. M. and Alves, H. (2017). Sustainable improvement of competitiveness in rural tourism destinations: The quest for tourist loyalty in Spain. *Journal of Destination Marketing and Management*, 6(3), pp. 252–266.

Carayannis, E. G., Ferreira, F. A. F., Bento, P., Ferreira, J. J. M., Jalali, M. S. and Fernandes, B. M. Q. (2018). Developing a socio-technical evaluation index for tourist destination competitiveness using cognitive mapping and MCDA. *Technological Forecasting and Social Change*, 131, pp. 147–158.

Cetin, G. and Bilgihan, A. (2016). Components of cultural tourists' experiences in destinations. *Current Issues in Tourism*, 19(2), pp. 137–154.

Chatkaewnapanon, Y. and Kelly, J. M. (2019). Community arts as an inclusive methodology for sustainable tourism development. *Journal of Place Management and Development*, 12(3), pp. 365–390.

Choi, S., Kim, J.-N. and Cai, L. A. (2012). T*ourism Loyalty: An Extended Communicative Action Model*. West Lafayette: Purdue Tourism and Hospitality Research Center.

Cong, L. C. (2016). A formative model of the relationship between destination quality, tourist satisfaction and intentional loyalty: An empirical test in Vietnam. *Journal of Hospitality and Tourism Management*, 26, pp. 50–62.

Darbari, J. D., Kannan, D., Agarwal, V. and Jha, P. C. (2019). Fuzzy criteria programming approach for optimising the TBL performance of closed loop supply chain network design problem. *Annals of Operations Research*, 273(1–2), pp. 693–738.

Dobrovolskienė, N., Tamošiūnienė, R., Banaitis, A., Ferreira, F. A. F., Banaitienė, N., Taujanskaitė, K. and Meidutė-Kavaliauskienė, I. (2019). Developing a composite sustainability index for real estate projects using multiple criteria decision making. *Operational Research*, 19(3), pp. 617–635.

Estêvão, R. S. G., Ferreira, F. A. F., Rosa, Á. A., Govindan, K. and Meidutė-Kavaliauskienė, I. (2019). A socio-technical approach to the assessment of sustainable tourism: Adding value with a comprehensive process-oriented framework. *Journal of Cleaner Production*, 236, 117–487.

Fernandes, I. D. S., Ferreira, F. A. F., Bento, P., Jalali, M. S. and António, N. J. S. (2018). Assessing sustainable development in urban areas using cognitive mapping and MCDA. *International Journal of Sustainable Development and World Ecology*, 25(3), pp. 216–226.

Fitra, A. and Leksmono, S. M. (2001). Pengembangan Kepariwisataan Berkelanjutan. *Jurnal Ilmu Pariwisata*, 6(1), pp. 87–88.

Foster, W. M., Coraiola, D. M., Suddaby, R., Kroezen, J. and Chandler, D. (2017). The strategic use of historical narratives: A theoretical framework. *Business History*, 59(8), pp. 1176–1200.

Franzoni, S. (2015). Measuring the sustainability performance of the tourism sector. *Tourism Management Perspectives*, 16, pp. 22–27.

Gallen, S. (2020). *The Response of Tourism Businesses vis-à-vis the Economic Ramifications of SARS-CoV-2. AIEST.* Retrieved from: https://www.aiest.org/news/ (accessed: 10th August, 2021).

Giglio, V. J., Luiz, O. J., Chadwick, N. E. and Ferreira, C. E. L. (2018). Using an educational video-briefing to mitigate the ecological impacts of scuba diving. *Journal of Sustainable Tourism*, 26(5), pp. 782–797.

Govindan, K., Shankar, M. and Kannan, D. (2018). Supplier selection based on corporate social responsibility practices. *International Journal of Production Economics*, 200, pp. 353–379.

Guzman-Parra, V. F., Quintana-García, C., Benavides-Velasco, C. A. and Vila-Oblitas, J. R. (2015). Trends and seasonal variation of tourist demand in Spain: The role of rural tourism. *Tourism Management Perspectives*, 16, pp. 123–128.

Hall, C. and Page, S. (2014). *The Geography of Tourism and Recreation: Environment, place and Space* (4th ed.). New York: Routledge.

Hao, F., Xiao, Q. and Chon, K. (2020). COVID-19 and China's hotel industry: Impacts, a disaster management framework and post-pandemic agenda. *International Journal of Hospitality Management*, 90, p. 102636.

Hashemi, N. and Ghaffary, G. (2017). A proposed sustainable rural development index (SRDI): Lessons from Hajij village, Iran. *Tourism Management*, 59, pp. 130–138.

He, P., He, Y. and Xu, F. (2018). Evolutionary analysis of sustainable tourism. *Annals of Tourism Research*, 69, pp. 76–89.

Heslinga, J., Groote, P. and Vanclay, F. (2018). Understanding the historical institutional context by using content analysis of local policy and planning documents: Assessing the interactions between tourism and landscape on the Island of Terschelling in the Wadden Sea Region. *Tourism Management*, 66, pp. 180–190.

Huang, C. (2020). *Coronavirus: For China, the Economic Pain Has Only Just Begun.* Retrieved from: https://www.scmp.com/week-asia/opinion/article/3076175/coronavirus-china-economic-pain-has-only-just-begun (accessed: 20th September, 2021).

Ismayanti. (2013). *Pengantar Pariwisata.* Jakarta: Grasindo.

Jader, R. A. Al. and Santosa, I. (2015). A conceptual development on the mediating role of E-service recovery on the relationship between customer determinants and customer retentions in the airline industry in Malaysia (a structural equation modelling approach). *Indian Journal of Commerce and Management Studies*, VI(1), pp. 103–109.

Jones, P., Hillier, D. and Comfort, D. (2016). Sustainability in the hospitality industry: Some personal reflections on corporate challenges and research agendas. *International Journal of Contemporary Hospitality Management*, 28(1), pp. 36–67.

Kernel, P. (2005). Creating and implementing a model for sustainable development in tourism enterprises. *Journal of Cleaner Production*, 13(2), pp. 151–164.

Li, Y., Zhang, H., Zhang, D. and Abrahams, R. (2019). Mediating urban transition through rural tourism. *Annals of Tourism Research*, 75, pp. 152–164.

Loi, C., Papatheodorou, A. and Glyptou, K. (2014). Tourism and the application of environmental legislation: The case of Rhodes, Greece. *Tourism Today*, Fall, pp. 143–163.

Maclean, M., Harvey, C., Suddaby, R. and O'Gorman, K. (2018). Political ideology and the discursive construction of the multinational hotel industry. *Human Relations*, 71(6), pp. 766–795.

Mataveli, M. and Gil, A. J. (2018). Loyalty to rural tourism in Brazil: The moderating effect of the purchase frequency. In F. Ali, S. M. Rasoolimanesh and C. Cobanoglu (eds.), *Applying Partial Least Squares in Tourism and Hospitality Research*. Bingley: Emerald Publishing Limited, Bingley, pp. 109–127.

Muljadi, A. (2012). *Kepariwisataan dan Perjalanan*. Jakarta: Raja Grafindo Persada.

Murdy, S., Alexander, M. and Bryce, D. (2018). What pulls ancestral tourists 'home'? An analysis of ancestral tourist motivations. *Tourism Management*, 64, pp. 13–19.

Okezone.com (2021). *Sandiaga Uno & Hary Tanoesoedibjo Kolaborasi Percepat Kebangkitan Parekraf*. Retrieved from: https://travel.okezone.com/read/2021/02/26/406/2369131/sandiaga-uno-hary-tanoesoedibjo-kolaborasi-percepat-kebangkitan-parekraf (accessed: 20th September, 2021).

Osman, Z. and Sentosa, I. (2013). A study of mediating effect of trust on customer satisfaction and customer loyalty relationship in Malaysian rural tourism. *European Journal of Tourism Research*, 6(2), pp. 192–206.

Pérez Gálvez, J. C., Granda, M. J., López-Guzmán, T. and Coronel, J. R. (2017). Local gastronomy, culture and tourism sustainable cities: The behavior of the American tourist. *Sustainable Cities and Society*, 32, pp. 604–612.

Purnomo, S., Rahayu, E. S., Riani, A. L., Suminah, S. and Udin, U. (2020). Empowerment model for sustainable tourism village in an emerging country. *Journal of Asian Finance, Economics and Business*, 7(2), pp. 261–270.

Putra, A. M. (2006). Konsep Desa Wisata. *Jurnal Manajemen Pariwisata*, 5(1).

Rajaratnam, S. D., Nair, V., Sharif, S. P. and Munikrishnan, U. T. (2015). Destination quality and tourists' behavioural intentions: Rural tourist destinations in Malaysia. *Worldwide Hospitality and Tourism Themes*, 7(5), pp. 463–472.

Reis, I. F. C., Ferreira, F. A. F., Meidutė-Kavaliauskienė, I., Govindan, K., Fang, W. and Falcão, P. F. (2019). An evaluation thermometer for assessing city sustainability and livability. *Sustainable Cities and Society*, 47, p. 101449.

Romão, J., Machino, K. and Nijkamp, P. (2018). Integrative diversification of wellness tourism services in rural areas: An operational framework model applied to east Hokkaido (Japan). *Asia Pacific Journal of Tourism Research*, 23(7), pp. 734–746.

Sangpikul, A. (2018). The effects of travel experience dimensions on tourist satisfaction and destination loyalty: The case of an island destination. *International Journal of Culture, Tourism and Hospitality Research*, 12(1), pp. 106–123.

Sastrayuda, G. S. (2010). *Konsep Pengembangan Kawasan Agrowisata*. Retrieved from: http://file.upi.edu.gumelar_s.go.id (accessed: 10th July, 2021).

Shankar, K. M., Kannan, D. and Kumar, P. U. (2017). Analyzing sustainable manufacturing practices: A case study in Indian context. *Journal of Cleaner Production*, 164, pp. 1332–1343.

Suhartanto, D., Brien, A., Primiana, I., Wibisono, N. and Triyuni, N. N. (2020). Tourist loyalty in creative tourism: The role of experience quality, value, satisfaction and motivation. *Current Issues in Tourism*, 23(7), pp. 867–879.

World Commission on Environment and Development (WCED) (1987). *Report of the World Commission on Environment and Development: Our Common Future*. Retrieved from: http://www.un-documents.net/our-common-future.pdf (accessed: 20th September, 2021), p. 2.

Weaver, D. B. and Jin, X. (2016). Compassion as a neglected motivator for sustainable tourism. *Journal of Sustainable Tourism*, 24(5), pp. 657–672.
World Travel & Tourism Council (WTTC) (2020). *Travel & Tourism: Global Economic Impact & Issues 2021.* Retrieved from: https://wttc.org/Research/Economic-Impact/Indonesia.pdf (accessed: 20th September, 2021).
Žabkar, V., Brenčič, M. M. and Dmitrović, T. (2010). Modelling perceived quality, visitor satisfaction and behavioural intentions at the destination level. *Tourism Management*, 31(4), pp. 537–546.

Part 4

COVID-19 and Post-COVID-19 Challenges and Implications

12 Tourism Business in the COVID-19 Pandemic

Tourism Industry Technology Framework (TITEF)

Emi Normalina Omar, Nur Atiqah Rochin Demong, and Siti Noorsuriani Maon

Introduction

The tourism industry has become more popular over recent years, with many people taking leisure time and participating in it. Tourism is a combination of processes, activities, and services involved when people travel to other locations, either domestically or internationally, for business, leisure, or personal reasons, individually or in a group. Tourism-related industries include hospitality, transportation, travel facilitation and information, attractions, and entertainment, such as hotels, accommodation, restaurants, airlines or car rental, travel agents, tourist information centres, shopping malls, and national wildlife parks. In Malaysia, the healthcare, education, and agriculture sectors also play crucial roles in the tourism industry (Ministry of Tourism and Culture Malaysia, 2020). These tourism sectors have used technologies to assist in the growth of tourism businesses as well as boosting the tourism industry. According to Stipanuk (1993), technology has several significant roles in the massive economic growth of the tourism business, including as creator, protector, enhancer, focal point, and destroyer of tourism experience and as a powerful tool of the tourism industry. The impact that technology has had on other sectors within the tourism industry is proven and also give multiplier effects to other sectors. The evolution of technology in tourism industries is apparent in the outcomes such as the evolution of player strategies and scheduling, the absence of physical tourist attractions, and the impact on overall visitor satisfaction. For example, innumerable digital tools have made our travelling experience efficient and pleasant, from online booking platforms to luggage trackers.

Moreover, the impact of technological advancement in tourism has created a travelling experience that is efficient and pleasurable. However, the COVID-19 pandemic has reshaped the tourism business. It has opened new opportunities for technology to evolve and create safer experiences for tourists, travel hubs, and the people who work in the industry.

The purpose of this chapter is to introduce the "technology framework" of the tourism industry, explore the most important concepts, deal with information technology, especially the Internet of things (IoT) and other technologies,

DOI: 10.4324/9781003315179-17

and draw conclusions on usage issues and challenges. This paper aims to propose a framework for tourism industry information using the Tourism Industry Technology Framework (TITEF) model.

Business Tourism Industry

The tourism industry offers numerous benefits to society by providing a cultural, socio-economic, and economic boost. In turn, tourism creates opportunities for job creation, increases a city's revenue, and improves a city's infrastructure through improvements and renovations. However, most businesses were affected by the COVID-19 pandemic outbreak, particularly the food supply chain, which saw a mismatch of supply and demand (Khan et al., 2021). This was also worsened by the weak business structure and low food supply chain resilience (Sharma et al., 2020). Therefore, most businesses are trying to maintain a consistent flow of essential goods and services as the globe strives to navigate this challenging moment (Sharma et al., 2020).

Hence, the business owner has made numerous efforts to survive during COVID-19. Many business owners across the country have been actively doing whatever they can to stay in business during COVID-19. The phase of "Transition to Endemic" has recently begun in Malaysia.and expected to endure until 2020. The effort also continues to the year 2021. As the crisis develops, there is rising worry about the pandemic's economic and societal consequences, which may establish standards for the "new normal" (Sharma et al., 2020). Eventhough the business's survival during the COVID-19 is really low, however, some of the companies remained able to retain their competitiveness throughout COVID-19. The best way to survive during COVID-19 is to focus on research and business development instead of promotion and content production.

Impact of COVID-19 on the Tourism Industry

There are negative and positive impacts of COVID-19 on the tourism industry. In the event of a pandemic, people may be cautious about travel. The tourism industry can greatly impact the number of visitors who visit your locality or state, reducing the visitation numbers. When it comes to the number of visitors to the region tourism has the ability to have an impact on national statistics, which may contribute to beneficial results. The more visitors that come to your locality or state, the more likely it is that tourism will eventually have the power to do more good or cause more negative consequences. There are many ways that tourism is effective in many aspects of your locality, state, or country. However, it's important to keep in mind that tourism significantly influences the country's economy.

According to the Malaysia National Tourism Policy 2020–2030, the main aim of the policy is to formulate a National Tourism Policy based on principles of competitiveness, sustainability, and inclusiveness to place Malaysia as a global top ten tourism destination in both arrivals and receipts (Ministry of Tourism and Culture Malaysia, 2020).

Six transformation strategies have been listed as follows:

- Strengthen governance capacity
- Create special tourism investment zones
- Embrace smart tourism
- Enhance demand sophistication
- Practice sustainable and responsible tourism
- Upskill human capital

These six transformation strategies show the areas that need to be enhanced and developed for sustainable tourism. Sustainable tourism is an approach to tourism that conserves natural and cultural resources. Tourism in the long term can be maintained with sustainable tourism. Sustainable tourism is a great way to consider the economic, social, and environmental impacts on a destination and to understand the needs of both visitors and the destination industry.

COVID-19 has a huge impact on business in every industry sector, including electricity, agriculture, retailing, warehouse, food transport, logistics, pharmaceuticals, and the tourism industry globally. The great impact of COVID-19 on the tourism industry can be seen in the number of international tourists for Q1 2020; there was a 22% decline as compared to the previous year 2019, and it is expected to decline by 60–80% for the whole year due to the restriction of travel, quarantines, and transportation closures (American Hotel & Lodging Association, 2021).

Technologies Used in the Tourism Industry

Today, the majority of tourist services and goods are purchased online. Customers obtain lower prices when they book online, and it is easier for them to check for information and other details about the services on the web. Therefore, online booking has risen in recent years. Customers are drawn to the incentives offered by suppliers for electronic purchases. There is a lack of product availability in local markets, a larger assortment of goods and services, lower prices, and referrals from friends. The technology uses digital tools such as cellular telephones, computers, and smartphones to perform business functions and processes like spreadsheets for budgeting. For the most part, the technology system is fast spreading across the tourist sector, and no participant will be able to avoid technology's effects in a significant way. Technology includes the entire range of electronic tools that facilitate the operational and strategic management of organizations by enabling them to manage their information, functions, and processes as well as to generally communicate interactively with their stakeholders to achieve their objectives, demonstrating how technology includes the entire range of electronic tools that facilitate the operational and strategic management of organizations.

Technologies provide online reservation and booking systems to match their customers' expectations. Some of the advantages of technology in the tourism industry are improvements in internal efficiency and capacity and yield management, more effective engagement with customers, and product and service customization. Technology increases the point of sale and revolutionizes tourist intermediation, allows visitors to speak with one another, encourages effective collaboration among system partners that values all parties or interests, and increases operational and geographic breadth by providing strategic tools for global expansion.

Revolutions of Technology in the Tourism Industry

The hospitality business is a key component of the tourist industry. Every overnight stay during any vacation or visit necessitates hospitality sector services. This tourism business sector is the second most important point of contact for clients after the travel sector. The rapid growth of communication technology has greatly aided the industry's development. Tourists may utilize technology to make personal selections about travel subsidiaries, lodging, cuisine, and activities available at their destination while sitting at home. Technology has made the travel and tourism industry more efficient, cost-effective, rewarding, and customer-focused in a major way. In the tourism business, technology plays a significant role in service delivery.

To some extent, the industry is attempting to use smartphones for future communication purposes. Because services are becoming commodities, every service provider is looking for new values to add to the services they give. Providers use enhanced products and services to develop exceptional service. The majority of augmentation products and services are generated from technology usage.

e-Tourism

The advancement of information technologies has had a strong influence on marketing tourism, advertising, and sales in recent years. With the introduction of recent technological advances, a novel kind of tourism, e-tourism, has emerged that has fundamentally reshaped the tourism business, particularly during the COVID-19 pandemic. People are increasingly turning to the Internet to get

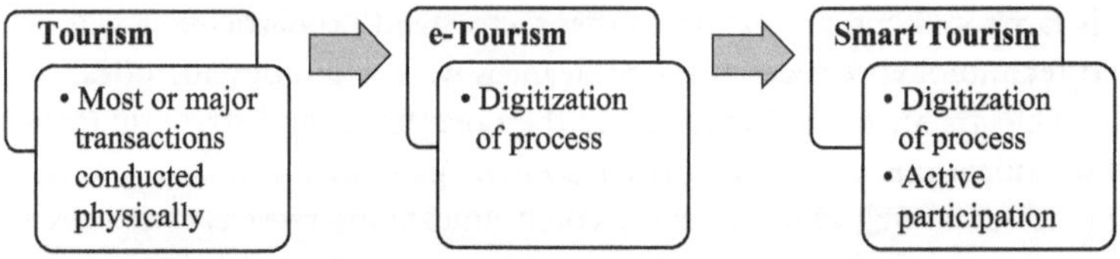

Figure 12.1 Revolution of technology in tourism industry (source: developed by the authors, 2021).

information about tourist sites and deals. Statistics show that Internet purchases are becoming increasingly popular in many countries. There is no better way for the travel and tourism business to flourish worldwide than via e-tourism. Because the development of e-tourism is dependent on information and communication technology, information and communication technology (ICT) plays a critical role in the worldwide expansion of tourism, and the use of software in ICTs is required. Although e-tourism is new to society, the concept has been causing joy within people's hearts as it offers an array of new experiences.

The e-Tourism Marketplace

Information and communication technologies have had a rising influence on tourist promotion, marketing, and sales in recent years. With the introduction of new forms of technology, a novel kind of tourism, e-tourism, has emerged that has profoundly revolutionized the tourism business, particularly amid COVID-19 attacks or pandemics. People are increasingly turning to the Internet to get information about tourist sites and deals. The sites should include general information on the region, such as map data, passports, and weather, and regional knowledge about travel agencies such as private interest trips, a list of major attractions and craft centres, national or international parks, business, vacation rentals, food outlets, tourism boards, air carriers, and transportation, practical information about adventures, guidance, interpreters, doctors, and car rental agencies, and local craft information (related organizations and associations, related trades, items, and so forth).

Smart Tourism

There has been an expansion of technology worldwide with regard to tourism. In recent years, tourism has seen a significant shift, made possible by integrating such technologies as the IoT, wearables, augmented reality, and augmented manufacturing. The global proliferation of technology has expanded the concept of tourism through a wide range of emerging ideas that the industry is keen to capitalize on. Transforming tourism means reaching an advanced stage in development. Smart tourism, advanced tourism development, and the new and current era are used to describe the new, more sophisticated movement and the perspective taken by the tourism industry. Smart tourism is a new form of tourism development in the current era, offering a variety of new opportunities for growing and evolving the industry to consider the latest trends and developments. It is an application of technology which integrates different aspects of tourism such as travel, accommodation, and activities.

The primary aim of smart tourism is to improve resource management efficiency, enhance the experiences of various tourists, gain a competitive edge, and sustain the industry for the future by promoting sustainability through technological innovation and practices. According to Zhang and Yang (2016), smart tourism has been affected by the advancement of information technologies and is emerging as a promising strategy to deal with dynamic market conditions. Smart tourism technology consists of three main components: cloud services, the IoT, and information communication technology, including mobile communication technology and artificial intelligence. It is an integrated application and innovation of technology and is mainly used in positioning, guiding, touring, and booking.

Generally, the concept of "smart tourism" was inspired by IBM's "Smarter Planet" and "Smarter City." The five most common methods of implementing smart tourism are (1) smart tourism accessibility options, (2) smart tourism information sharing, (3) smart tourism sustainability initiatives, (4) smart tourism research and management tools, and (5) smart tourism related to the tourist experience.

Tourism Industry Technology Framework

In this chapter, as seen in Figure 12.2, we define the "technology framework" for the tourism industry, go over the top concepts involved, delve into the information technology, specifically the IoT, and other technologies, finally it also discuss the issues and challenges of using technology in the tourism industry. This paper proposes a framework of tourism industry information available through a model called the TITEF, which offers a clear insight into the various concepts, information technology, business application, development process, and issues or challenges of using technology in the tourism industry. The relationship between these five areas is when engineering is done, the concepts are developed, the

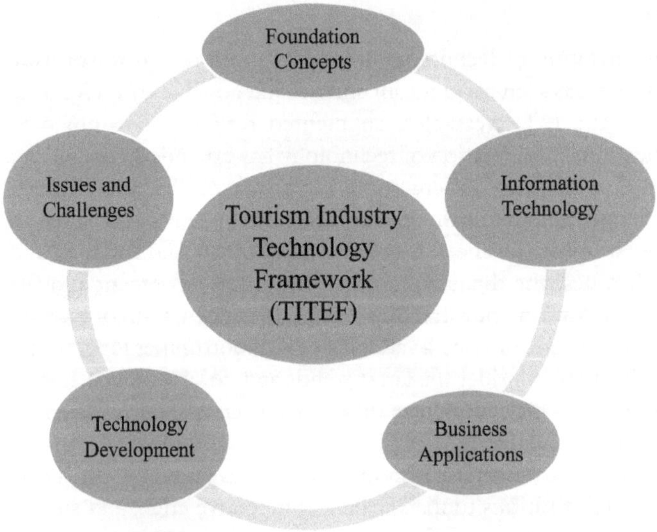

Figure 12.2 Tourism Industry Technology Framework (source: developed by the authors).

technology is tested, the application of those concepts in business is implemented, and the development process is looked at. The process's challenges and setbacks are posed here to help meet your tourism industry needs.

Foundation Concepts

A touristic technology industry framework has the mission of enabling the world to embrace tourism. Foundation concepts encompass the main components, main activities, and an overview of the project. The hierarchy flows down to the fundamentals of tourism and the essential software, hardware, data, strategies, and platform used in the industry, which can be studied in detail as can their respective roles. People are an important factor in decision making and strategic planning for the tourism industry and technology to complete transactions. Another important component of any tourist industry is the hardware that used in the respective businesses to make it as efficient as possible. Hardware generally refers to any devices or peripherals used in the design and development process for computing technologies to produce outputs, such as the computer's processor, computer memory, and input devices such as a keyboard to input characters, a microphone to input voice, and a camera to input videos. Another component in this area is that different devices have other functions to help gather data types and process them using the software. Software is applications specifically designed to process information and create new outputs for the users of a system or platform, and it is either used to convey information or to solve a problem. Another component will be data that refer to variables that a user can input. Finally, there is the network used to distribute the data or information. The focus for the network will be related to issues of what is or will be the best option for data or information distribution. Additionally, there is the choice of the network that is to be used to distribute, or the layout or the intermediary utilized to transmit, the data or the information presented.

Another one of the foundation concepts of this area will be the main activities involved in technology in the tourism industry. An important element of exploring the foundations of technology in the tourism industry will be exploring how many key activities occur in the technology sector. There are four key activities in the tourism industry: collection, input, input transform, and output for collection, which is the process of capturing information about tourists and the location of their activities. The next is input, which is the process of extracting data for transformation. To understand how this subpart works, one must consider that the data transformation process includes some or all of the data collected for decision making or strategic planning, either in the short term or long term. Input transforms need to be done for reporting and analysis that will help the users, specifically the customers and organizations' stakeholders, meet their requirements of using the technology in the tourism industry. The result of the reporting and analysis can be represented using the dashboard or other tools for presentation will be the output of the main activities conducted. Strategies and platforms have been implemented in the tourism industry to satisfy users' needs and requirements. Different strategies and platforms are being used in the tourism industry. Tourism firms often succeed at fulfilling their end of the bargain and pleasing their users by employing strategies and platforms.

Information Technology in the Tourism Industry

Data

Data come in many forms. Data types include numbers or integers, alphabetical symbols, voice, image, or video that can be categorized into structured, semi-structured, or unstructured data types for different purposes. All these data will be converted into digital form for processing purposes in technology and uploaded to the Internet. It is imperative to employ a systematic approach to ensuring data recorded or gathered are valid, updated, and united.

Hardware

Hardware includes any devices and medium that helps collect the info, process it, and produce the output for a specific purpose. Hardware can be categorized into three main categories: input, process, and output devices, which is essentially quite significant. Input devices used in the tourism industry, such as speech recognition with new tools available in the market, are substantial. There are many devices for processing and output, and most of these devices are user-friendly, contrary to popular belief.

Software

Software refers to any programmes used to process the input and generally produce the output and will be used together with hardware, which is significant. The fast-changing trends in consumer behaviour and technology have created new markets and opportunities for new software development, which is quite significant. In the tourism industry, intangible tourist services and goods require a demonstration of technology that allows them to be seen by the consumer, either in written documents or audio and video formats, referring to the output from software and hardware that receive data as input in the system. Implementing technology becomes critical, and software used to retain information as needed happens. Every organization in the tourist sector employs various types of software designed to meet the requirement for information processing for their commercial purposes.

Network

A network is a mechanism for distributing or sharing data or information across all players in the tourist sector. The worldwide development of tourism has altered the way customer services are provided, and the use of advanced technology has changed the demand and requirement of consumers in modern society, particularly in conveying data or information for various objectives. Another aspect of the learning was the amount of comprehension at each hierarchy level inside the organization. Surveys and interviews were intriguing experiences that taught us how and why information must be gathered at various levels to assess the service process and the product's strengths and weaknesses, and the features it contains. Distinct types of networks have varied requirements and functions and different costs.

Business Applications

This area of TITEF will explain sectors involved in the tourism industry related to the use of technology and the criteria needed to complete certain functions to ensure the transaction runs smoothly. This paper will focus on four major functional areas related to the tourism industry: marketing, applied in major sectors such as transportation, accommodation, food and beverages, and entertainment. Table 12.1 depicts the business function and technologies related to the tourism industry.

Business Functions and Technology Applications in the Tourism Industry

Technology is deeply ingrained in the tourism business, influencing every action in the market. The travel industry faces both opportunities and problems in reaching its clients.

Table 12.1 Depicts the business function and technologies related to the tourism industry

Business roles and functions	*Technology applications*

• Selecting travel destination
• Reservation/booking and
  • Transportation
  • Accommodation
  • Food and beverages
  • Entertainment
  • Services
• Travel insurance
• Payment
• Review and rating
	• Geospatial information technologies
	• Augmented reality (AR) for tourist destinations, gamification, city tours
	• e-marketplace for tourism (marketplaces business model) for
	• Check-in kiosk, keyless mobile entry, voice assistance room service, websites, email, phone calls, transit and baggage alerts, augmented reality for user experience, voice recognition, and fingerprint recognition
	• Social media applications such as Facebook, Twitter, Instagram, WhatsApp, etc.
	• Online travel agencies such as Agoda.com, booking.com, Airbnb, and many more

(Source: developed by the authors.)

Technology Development

Several voice-search mechanisms have contributed to the increasing significance of voice search for travel technology. Mobile phones, smart speakers, and artificial intelligence (AI) assistants have all contributed to the shift in travel technology. A rising number of customers, particularly, are searching for and buying airline tickets, hotel rooms, and travel experiences by utilizing voice search, finding and buying what they need and desire at the click of a fingertip. Travel agencies, in particular, are now using voice search to help find and purchase airline tickets, hotel rooms, and travel packages online as a rising number of clients request these services. The smartphone is undoubtedly the primary character in the new modes of transportation and has evolved into a tour guide and travel agency, restaurant locator, map, and much more. It is present throughout the entire travel purchasing process. Many travel agents utilize apps to connect with travellers who need to reschedule.

Furthermore, there are also resort-specific smartphone applications that allow visitors to raise concerns about management. There are many different definitions for travel technology, but it can refer to information technology (IT), e-commerce, and related technology in industries such as transportation, logistics, and tourism. Hotel and tourism technology is designed to benefit the companies and their staff by improving client experiences, helping with the planning and administration of travel, or both. Finally, automating travel and related procedures, saving time, lowering costs, and providing resources in an ever-changing array are just some of the many reasons why so many companies have put travel technology (applied or utilized) to providing customers with a more seamless travel experience. Technology can improve the consumer experience in several ways, including before the trip, during, and after. Numerous organizations may soon be able to figure out the most effective way for their travel technology to automate operations that are otherwise too human-intensive for a typical business. Businesses can potentially decrease workforce needs, lower expenses, and increase revenue with improved financial outcomes.

Issues and Challenges

Industry Roles

The economic challenges and responses of the tourism industry during the COVID-19 pandemic and how it will impact the country were determined by a study looking at the economy. If the economy of the particular country improves, it shows that the country can survive. The industry players and the country benefitted from this positive impact. On the other hand, if there is an economic downturn or the economic situation turns out negatively for the country, then the country will suffer a negative effect. The negative impact needs to be addressed to maintain the country's economy and well-being. The tourism business industry has been devastated through the closing of the border by the government. When the government shuts down the border, the tourism business or industry cannot survive. This scenario is worst when it involves a serious increase in retrenchment, job loss, and high unemployment due to COVID-19. Therefore, a new concept in the tourism has shifted from physical tourism into the digital space through developments in technology. Through the technological advancements, the concept of tourism has changed from traditional physical travel to online travel. The industry is in the midst of a paradigm shift that gives new experiences to tourists through online known as "smart tourism". There are new opportunities for people who used to travel but are now staying at home due to the epidemic to engage in "smart tourism". Therefore, the tourism industry needs to adapt to the current worldwide pandemic, which will lead to a market explosion in new business for them to continue to be competitive or enhance their offerings.

Government Roles

The policy implemented by the government due to COVID-19 in regard to tourism is as follows:

a) Travel Restrictions
Due to global pandemic outbreaks and a travel restriction imposed by the government, an announcement was made regarding the government's support for the worldwide economic recovery through the offering of bubble travel. Because of the pandemic, which has led to travel restrictions, the government has decided to implement a complete bubble travel for the entire country to boost the global economy. The government has chosen to establish a travel bubble for the entire country due to the pandemic in order to strengthen the country's economy. Indeed, in response to the pandemic, the World Health Organization has imposed travel restrictions to recover the country's economy. The government has announced a bubble travel plan for the whole population.

b) Border Close
The government's announcement of closing the border to tourism has put the industry at great risk. This scenario may weaken the tourism industry if the border closes. Since the tourism sector cannot continue its operation if the border closes, this scenario is extremely giving high risk to other industries too. Consequently, the government's decision in closing of the border has made it impossible for the tourism industry to survive. The tourism industry cannot survive from a structural perspective, meaning that the government shutdown shutters the gateway for tourism funding. This means that neither the global nor domestic tourism businesses could survive if governments closed the borders.

c) Incentives
Government incentives can boost the industry players, who will be faced with rebuilding their tourism business over the next several months. The industry players can recover and rebuild their businesses with the incentives given

by the government. The removal of travel restrictions by the government may offer relief to the country's economy. As a result, the operators and industry players must follow the new regulations' operating hours' policy. The removal of travel restrictions, open job opportunities, and a recovering economy put this country back on the map.

d) Vaccine for People
Since the government has imposed a new policy that people must have two doses of vaccine, people now feel safe enough to eat and enter shopping malls. This makes people feel safer than they used to. Before entering the shop, a customer should check their body temperature and have their completed vaccination certificate in hand. This ensures that the customer is healthy and not spreading disease to other customers.

Conclusion

This chapter proposes a framework of tourism industry information available through a model called the TITEF that offers a clear insight into the various concepts, information technology, business application, development process, and issues or challenges of using technology in the tourism industry. A touristic technology industry framework has the mission of enabling the world to embrace tourism. Foundation concepts encompass the main components, main activities, and an overview of the project. The hierarchy flows down to the fundamentals of tourism and the essential software, hardware, data, strategies, and platform used. The tourism industry has four key activities: collection, input, input transform, and output. The result of the reporting and analysis can be represented using the dashboard or other tools for presentation. Tourism firms often succeed at fulfilling their end of the bargain by employing strategies and platforms.

Data come in many forms, including numbers, alphabetical symbols, voice, image, or video, categorized into structured, semi-structured, or unstructured data types. All these data will be converted into digital form for processing purposes in technology and uploaded to the Internet. Hardware includes any devices or medium that helps collect the info, process it, and produce the output for a specific purpose. Hardware can be categorized into three main categories: input, process, and output device. There are many devices for processing and output devices, and most of these devices are user friendly. Every organization in the tourist sector employs various types of software designed to meet the requirements of information processing for their commercial purposes. Fast-changing trends in consumer behaviour and technology have created new markets and opportunities for new software development, which is quite significant. A network is a mechanism for distributing or sharing data or information across all players in the tourist sector. Distinct types of networks have varied requirements and functions and different costs. Another aspect of learning is the amount of comprehension at each hierarchy level inside the organization. Voice search is increasingly used to find and buy airline tickets, hotel rooms, and travel packages online. Mobile phones, smart speakers, and AI assistants have all contributed to the shift in travel technology. Many travel agents utilize apps to connect with travellers who need to reschedule.

References

American Hotel & Lodging Association (AHLA). (2021). *Covid-19's Impact on the Hotel Industry*. Retrieved from: https://www.ahla.com/covid-19s-impact-hotel-industry (accessed: 10th August, 2021).

Khan, S. A. R., Ponce, P., Tanveer, M., Aguirre-Padilla, N., Mahmood, H., and Shah, S. A. A. (2021). Technological innovation and circular economy practices: Business strategies to mitigate the effects of COVID-19. *Sustainability*, 13(15), pp. 1–17.

Ministry of Tourism and Culture Malaysia (2020). National tourism policy 2020-2030. In *Compensating for Wetland Losses Under the Clean Water Act*. Retrieved from: http://www.censusindia.gov.in/2011census/PCA/PCA_Highlights (accessed: 10th August, 2021).

Sharma, A., Adhikary, A. and Borah, S. B. (2020). Covid-19's impact on supply chain decisions: Strategic insights from NASDAQ 100 firms using Twitter data. *Journal of Business Research*, 117, pp. 443–449.

Stipanuk, D. M. (1993). Tourism and technology. Interactions and implications. *Tourism Management*, 14(4), pp. 267–278.

Zhang L. and Yang J. (2016) Smart tourism. In J. Jafari and H. Xiao (eds.), *Encyclopedia of Tourism*. Cham: Springer, https://doi.org/10.1007/978-3-319-01384-8_175

13 Contactless Hospitality Technology in the Post-COVID-19 Era

Fernando Mayor-Vitoria

Introduction

The health crisis that started in 2019, due to COVID-19, has made it clear that the hotel sector must transform itself and seek innovative solutions to face the problems arising from this crisis and other potential crises that may come. In this situation, the construction sector can collaborate with hotels to help them solve new problems, contributing to health care by designing new infrastructures that require creativity and innovation. This contribution is enriched if it is made as part of a Corporate Social Responsibility strategy.

In the tourism sector, environmental management and the minimization of the impacts derived from its activity have been the most common actions put in place by Corporate Social Responsibility during the last years of the 20th century. The specific measures developed by the sector have largely taken the form of eco-labelling and codes of conduct. Epler Wood and Leray (2005) also agree, pointing out that other aspects of Corporate Social Responsibility have been ignored. Welford, Ytterhus, and Eligh (1999) explain that these omissions are due to the fact that sociocultural problems were often specific to the locality where the business activity was located, and thus varied according to the destination. However, it is now clear that there are certain problems that have become globalized and require solutions that emanate from collaboration between different sectors and public administrations. Companies must consider what the social demands are and what level of contribution their company can make, especially when it operates in developing countries (Eweje, 2006), but in health crises such as the current one, these demands are globalized.

Actions that integrate Corporate Social Responsibility in the tourism sector must go beyond environmental management, although this is very important (Lund-Durlacher, 2015). An outstanding step in the development of Corporate Social Responsibility in tourism sector companies was achieved after the World Summit on Sustainable Development in 2002, when the tourism sustainability agenda changed and refocused on pro-poor tourism strategies and the Millennium Development Goals. Today, the 17 Sustainable Development Goals, projecting to 2030, follow this path (Schwarz, 2018).

DOI: 10.4324/9781003315179-18

The health crisis that began in 2019, due to COVID-19, has shown that the hotel sector must transform itself and seek innovative solutions to address the problems arising from this crisis and other potential crises that may come. In this situation, the construction sector can collaborate with hotels to help them solve new problems, contributing to health care by designing new infrastructures that require creativity and innovation. This contribution is enriched if it is made as part of a Corporate Social Responsibility strategy.

In addition to environmental aspects, a Corporate Social Responsibility strategy contributes to sustainable development through other types of actions related to human resource management, occupational health and safety, adaptation to change, development of local communities, and relations with other public and private actors, as recommended by the European Commission (Kuhlicke et al., 2011). Similarly, Coles, Fenclova, and Dinan (2013) understand that a Corporate Social Responsibility strategy in the tourism sector should be a voluntary strategy that should favour local communities, prioritizing sustainability, creating brand value, encouraging the participation of socially conscious tourists, attracting socially conscious investors, focusing recruitment on highly qualified human resources, and improving risk assessment and the capacity to respond to risks.

Nord (2007) points out that political decisions and the level of economic development of a region can affect the development of a Corporate Social Responsibility strategy and recommends that hotels should have homogeneous methods and units of measurement with the same overall reporting system that allows them to compare the results of their socially and environmentally responsible management and evaluate their performance in order to have evidence of the efficiency and success of their Corporate Social Responsibility strategy actions.

Numerous studies, such as Swarbrooke (1999), have focused on the social, ecological, cultural, and economic effects of tourism. Environmental and social impact studies have generally been carried out to assess the negative impacts of tourism on local communities and the physical environment, while studies focusing on the economic aspect generally evaluate by demonstrating its beneficial effects on income generation and the potential for job creation (Munasinghe, 1993).

From the perspective of Corporate Social Responsibility strategy actions, the aim is to promote studies that evaluate the positive impact that Corporate Social Responsibility strategy actions have both for the companies themselves and for society. Given the current situation, the hotel sector requires the construction sector to improve its activity and this, in turn, has an inherent social responsibility that materializes in a built environment that influences all aspects of life, especially human health, social behaviour, and cultural identity (Lu et al., 2016). Among the contributions that the construction sector can make to the hotel sector, those related to adaptation to change stand out (i.e., the introduction in hotels of innovations of various kinds that help the health of tourists and the population in general). This aspect would complement the hotel sector's commitment to sustainability (Horng et al., 2017).

Trends in Tourism Renovation before COVID-19

The global pandemic situation has given tourist accommodations the great challenge of adapting measures to guarantee and give confidence to customers, suppliers, and workers to comply with and maintain the protocols established by the different administrations. This chapter describes some measures that can be taken to adapt tourist accommodations to the new habits that have emerged in this pandemic and that are necessary to prevent the spread of COVID-19 and to generate confidence in all users.

In the first place, there is the redesign of wide, open, and transparent spaces in common areas. To this end, a study of mobility and internal circulation of customers in the tourist accommodations themselves should be carried out, considering aspects such as times at reception, for customer check-in and check-out, and how the waiting areas, elevators, bars, restaurants, meeting point areas, etc., are organized. The aim of this redesign is to avoid generating saturation in specific spaces and times, avoiding crowds so as not to produce a feeling of physical insecurity and to provide serenity in quiet accommodations for the enjoyment of clients' vacations.

A second measure, which would affect the design of the rooms, is the incorporation of small kitchens, so that the client can have independence in a very functional way. A proportion of clients demand to have the resources to prepare their own dishes in a simple way, with a microwave, a simple plate for cooking small meals, as well as a small refrigerator and a small sink, with basic kitchenware.

The characteristics of these kitchens could be adapted to different customer segments, such as families with children, seniors, and millennials. Another measure with a positive impact is the incorporation of new materials with straight lines that are easy to clean, with very little maintenance. Aesthetics and functionality should be prioritized in both indoor and outdoor materials, with simple lines, but very easy to maintain, as well as cleanability and resistance to use and weather, both indoors and outdoors. The above is related to the redecoration by means of simple, functional, and easy to clean furniture. It would be a question of incorporating interior and exterior furniture with very practical materials, which are easy to clean and maintain.

Establishments can also opt for the installation of light sensors to avoid pressing switches. This translates into the incorporation of people presence sensors activated by photoelectric cells and installed in areas such as common area corridors, landings, elevator waiting areas, restrooms, etc. In addition to eliminating the need to use switches, this measure has the added value of the cost savings that are generated, reducing electricity consumption at the time of passage or stay, which favours rapid returns on investment over time.

To avoid physical contact with the doors, automatic doors with presence sensors (already introduced in the market, but now generalized) could be adopted. This is a foretaste of what in the next generation will be called the Internet of things (IoT), which will become increasingly common in basic elements such as smart door openings. As they are connected to a control centre, they will open when a person is present, identifying him or her by means of voice keys, avoiding physical contact. A complementary measure to the above is the opening of doors by means of a contactless bracelet with a built-in contactless chip.

More and more lodgings are incorporating this type of element, obtaining advantages such as the opening of the room without a key or physical card, thanks to the incorporation of a microchip in the bracelet itself that coordinates with the sensor of the door of the room to open it. The tourist accommodation itself can take advantage of this measure to segment customers into the different types contracted: all inclusive, half board, or bed and breakfast according to the colour of the bracelet.

As it has a built-in chip, it can be programmed to pay for any consumption within the establishment. The wristband is protected and waterproof. Communication with customers would be improved by means of digital information panels using voice commands, which would avoid the need for touch devices in common areas. Therefore, voice-operated digital panels would be implemented to obtain information on accommodation, activities, external extra-hotel offerings, activity schedules, etc., connected to a database and the Internet. In addition, it would make it easier to adapt them to different languages and different types of customers.

Probably one of the biggest challenges is to modify the restaurants and dining areas of the accommodations, since they are one of the most important parts of the activity in tourist accommodations. Here, we propose the introduction of buffets and open-plan restaurants with just the right furniture (hot and cold), with the incorporation of natural light and the use of terraces with open spaces.

To avoid contact between customers at specific times without crowds, they would be managed by time slots, through pre-reservations between the customer and the accommodation, trying to cover the tastes and preferences of the customer.

For breakfasts, many establishments will opt for sober furniture with open spaces. Buffets should be short in length in order to avoid crowds and to ensure a smooth flow of customers.

Lunches and dinners should follow the same approach as breakfasts, but with more variety in both hot and cold dishes. The management of the establishment will be based on having agility, first, due to customer rotation and, second, to avoid saturation of customers at specific times.

For payment methods, contactless technology should be used, prioritizing smartphones as payment receptors for credit cards and mobile payments and avoiding cash payments. The first objective is not to have any contact with a material good, and the second is to reduce the time of money transfer between the lodging and the customer.

Post-COVID-19 Strategies and Technology

The COVID-19 pandemic has dragged the tourism sectors of all countries into an unprecedented situation. The reactivation of the sector from scratch is what differentiates this crisis from others that tourism has suffered for various reasons (terrorism, natural disasters, economic downturns). This raises many questions, despite the resilience it has shown in the past. Because of its link to the pandemic, the tourism environment questions relate to tourist behaviour and government policy, giving rise to various scenarios and consequences. The question regarding how the company will internalize these changes and react to them will be associated, singularly, with its organizational culture (Keyton, 2005).

In any case, it can be said that this crisis has two parts: the stage before the vaccine and the stage after the vaccine. Possibly, the first is the most delicate because of its connotations in terms of health security, having to live with the virus, and the risk of outbreaks without a vaccine, without specific drugs to treat the disease and without herd immunity.

Keeping in mind the memory of the tens of thousands of deaths and bearing in mind what a new escalation would mean, with the consequent return, it is time for individual responsibility to respect the protective measures dictated by the health authorities, for tour operators to adopt the protocols established by the latter, and for the public administration to balance, through empathy, active listening, and collaboration with the parties involved.

The duration of this phase will depend on the progress of scientific research, but everything seems to indicate that it will not be until 2022 that medicine will be able to allow us to enter the post-vaccine era, when confidence will be fully restored.

In this impasse, it is conceivable that, as a result of what we have learned, we will be better prepared to face both new waves of this coronavirus and other crises of this nature. It will also be a period in which society as a whole will have become accustomed to a new normality, with certain added precautions and new habits that will be incorporated into daily life and tourist activity.

But beyond the impact of the pandemic, the strategies of destinations and companies in the post-COVID-19 tourism market will have to take other factors into account. We must not lose sight of other fundamental vectors of change that, in conjunction with the legacy of this coronavirus, will converge in a profound reconfiguration of tourism activity. In particular, we must consider the technological revolution that is in the making and the adaptation to the challenges of climate change, including the application of the principles of the circular economy.

The COVID-19 pandemic, with the ensuing economic recession, leaves us with a scenario that forces us to manage scarcity. Indeed, it is no longer a question of managing abundance: everything has suddenly changed. Destinations will no longer fill themselves; we will have to make a much greater effort and bring all our capacities into play.

Especially in the period before the vaccine, shortages will be felt:

- To tourists (with an almost exclusively domestic market, and not all its segments fully operational) and spending (constrained by the economic recession). In other words, there will be a shortage of demand and then, services will presumably be subject to capacity limitations, as in the case of beaches.

- To information: pre-crisis references in terms of tourist behaviour patterns are no longer applicable.
- To confidence on the part of the issuing markets, undermined by the pandemic and the very serious impact it has had on Spain.
- To liquidity, with the sector's companies' financial muscle weakened by the forced hibernation of their activity.

Inbred behaviours will also be part of the legacy of COVID-19 in this phase as well as the announcement of redoubled public efforts in terms of promotion. In a market with a limited demand, quantitatively and qualitatively, the competition to attract customers will be desperate, with an evident risk of entering a spiral of falling prices, to at least generate movement and a certain sense of normality, although not profitability.

In a context in which mimicry will be stronger than usual, due to the unprecedented nature of this situation, the difference will not be so much in what is sold, but in how it is sold, which is why I advocate a new impetus in the tourism governance model that decisively includes and involves the people who live in the destination. Collaboration between public administrations and between these and business organizations should be strengthened with the people.

The recovery of tourism is everyone's business, not just that of public authorities and companies. Everyone is concerned and, therefore, we should all be called upon to make this cause our own. Tourism should not be built with its back to the people, but with the people. The greater the economic dependence on tourism, the more evident is the need to join forces. We must all assume the role of sellers of our destination. Its brand and slogan should be everyone's and should accompany us in our various activities, even if they are not directly linked to tourism. Just as a hotel receptionist is not only that, but also a salesman for his company, and a waiter in a restaurant is not only that, but also a salesman for his company, we residents of a destination should become salesmen for it. The support of all is needed to sell the destination in a context of exceptionally scarce financial resources, but also of leadership that motivates to do so.

Tourism, in most cases, lacks clear leadership, which is now, in situations of extreme difficulty, when it is most needed. If leaders emerge and are capable of energizing all the agents involved in the recovery, the possibilities of getting ahead will be greater, adding value and not lowering prices so much. During the shortage, we cannot afford any luxury.

Tourism, one of the major sectors affected by COVID-19, is extremely sensitive. That is why it is so important to transmit security and confidence. In order to plan a correct recovery, everything should be carried out in a rigorous manner, as it would be very negative if the tourist were to feel pressure or excessive control in the measures to be adopted.

Since the beginning of the pandemic, we have been able to analyse a large number of applications, services, and ICT solutions that could help minimize the effects of COVID-19. Faced with such an avalanche of initiatives, it is important not to lose objectivity. A part of the sector has even tried to protect itself and promote the concept of Smart Tourism Destinations as a solution to the problem, as it is a differentiating and competitive factor.

Let us recall one of the reference definitions of a Smart Tourist Destination:

> an innovative space consolidated on the basis of the territory and a state-of-the-art technological infrastructure. A territory committed to the environmental, cultural and socioeconomic factors of its habitat, equipped with an intelligence system that captures information in a procedural way, analyzes and understands events in real time, in order to facilitate the interaction of the visitor with the environment and the decision making of the destination managers, increasing its efficiency and substantially improving the quality of tourism experiences.
>
> (Medina and Plaza, 2015)

The concept highlighted the importance of innovation and sustainable development in its three areas: environmental, socioeconomic and cultural, and technology, to provide a better service to tourists. The concept did not refer to two fundamental aspects of governance, that is, residents who should benefit from the intelligent management of the territory in which they reside and accessibility.

In addition, the COVID-19 pandemic has increased awareness of the close interrelationships between environmental transformation and the emergence of new diseases (Rubin et al., 2020). Thus, the World Health Organization has been warning for some time that climate change may lead to the emergence of new epidemic diseases or increase their transmission, with the impacts we are seeing on the tourism sector (Mora et al., 2018).

In this context, in the field of tourism and cultural heritage, there are a series of specific objectives and lines of action related to the circular economy as one of the instruments for making progress in this adaptation to climate change, as it will help to reduce the impact of climate change (Scott, 2012).

Recommendations and Conclusion

The measures described are the main adaptations that customers are requesting in the accommodations to face the new tourism scenario. This is a problem that must be faced to adapt to a new situation with no defined time horizon, since these adaptations are made on the basis of health, administrative, and social criteria to prevent the spread of COVID-19.

The policymaker should have planned for better tourist accommodation procedures, such as physical distancing of at least two meters between people, the use of masks in indoor areas, hydro-alcoholic gels in almost all common areas, especially in bars, restaurants, and bathrooms, separating screens between customers and workers, separation between tables and chairs, as well as corridors for the flow of customers at specific points (nodes) and other minor measures. The proposals made in this chapter are measures, such as the opening of doors with microchip bracelets or information panels via voice command, which were planned for implementation in a period of five to ten years or even more, which would be implemented progressively more and more.

The pandemic has forced many lodgings to implement them now in order to adapt as quickly as possible and not lose competitiveness and business. The rest of the lodgings that cannot or will not adapt will cease to be attractive to an increasingly demanding market and, therefore, will disappear, closing their doors and turning into new buildings for other uses.

Since the beginning of the pandemic, technology has emerged strongly in the search for solutions of interest to many sectors, especially in tourism. Without downplaying the importance of such initiatives, it is clear that it is not proving to be as relevant as it may initially appear. The passion for technology can cloud the objectives or the effectiveness of the solutions. And for that same reason we cannot see blockchain or a set of apps as a way out, but we can see it as an effort for improvement.

Solving or minimizing the effects of COVID-19 is about rigor and compliance. We live in an era where we try to make everything smart: cities, whether major or minor, territories, islands, tourist destinations. As if it were a symptom of progress. Technology is not the end; it is the tool. Therefore, the way it is applied, its priority, and the wisdom to solve a problem will always be above any other factor.

The necessary interventions for the adaptation of hotels after the pandemic caused by COVID-19 will require, in some cases, refurbishment and maintenance. The material execution of these actions must be led by the agents of the construction sector, qualified people with the appropriate technical training. Among all the phases of the construction process, the design phase is key. Decisions that affect the building throughout its useful life, including the repercussions at the end of its useful life, must be taken at this stage, due to the repercussions involved and the additional costs involved in making these decisions at later stages.

The technical adaptations required by users will depend on the target audience, and have come up with some ideas. There are examples of good practices, but there is great resistance to change, and in this sense the good application of technology and the implementation of the circular economy are key.

References

Coles, T., Fenclova, E. and Dinan, C. (2013). Tourism and corporate social responsibility: A critical review and research agenda. *Tourism Management Perspectives*, 6, pp. 122–141.

Epler Wood, M. and Leray, T. (2005). *Corporate Responsibility and the Tourism Sector in Cambodia* (Vol. 34658). Washington, DC: The World Bank, p. 1.

Eweje, G. (2006). The role of MNEs in community development initiatives in developing countries: Corporate social responsibility at work in Nigeria and South Africa. *Business and Society*, 45(2), pp. 93–129.

Horng, J. S., Liu, C. H., Chou, S. F., Tsai, C. Y. and Chung, Y. C. (2017). From innovation to sustainability: Sustainability innovations of eco-friendly hotels in Taiwan. *International Journal of Hospitality Management*, 63, pp. 44–52.

Keyton, J. (2005). *Communication and Organizational Culture: A Key to Understanding Work Experiences.* Thousands Oak: SAGE Publications.

Kuhlicke, C., Steinführer, A., Begg, C., Bianchizza, C., Bründl, M., Buchecker, M., Buchecker, M., De Marchi, B., Di Masso, T. M., Hoeppner, C., Komac, B., Lemkow, L., Luther, J., McCarthy, S., Pellizzoni, L., Renn, O., Scolobig, A., Supramaniam, M., Tapsell, S., Wachinger, G., Walker, G., Whittle, R., Zorn, M. and Faulkner, H. (2011). Perspectives on social capacity building for natural hazards: Outlining an emerging field of research and practice in Europe. *Environmental Science and Policy*, 14(7), pp. 804–814.

Lu, W., Ye, M., Flanagan, R. and Ye, K. (2016). Corporate social responsibility disclosures in international construction business: Trends and prospects. *Journal of Construction Engineering and Management*, 142(1), p. 04015053.

Lund-Durlacher, D. (2015). Corporate social responsibility and tourism. In G. Moscardo and P. Benckendorff (Eds.), *Education for Sustainability in Tourism* (pp. 59–73). Berlin, Heidelberg: Springer.

Medina, A. C. and Plaza, A. G. (2015). El papel estratégico de las tecnologías de la información y las comunicaciones en el turismo. *International Journal of Information Systems and Software Engineering for Big Companies (IJISEBC)*, 2(2), pp. 52–69.

Mora, C., Spirandelli, D., Franklin, E. C., Lynham, J., Kantar, M. B., Miles, W., Smith, C. Z., Freel, K., Moy, J., Louis, L. V., Barba, E. W., Bettinger, K., Frazier, A. G., Colburn IX, J. F., Hanasaki, N., Hawkins, E., Hirabayashi, Y., Knorr, W., Little, C. M., Emanuel, K., Sheffield, J., Patz, J. A. and Hunter, C. L. (2018). Broad threat to humanity from cumulative climate hazards intensified by greenhouse gas emissions. *Nature Climate Change*, 8(12), pp. 1062–1071.

Munasinghe, M. (1993). *Environmental Economics and Sustainable Development* (Vol. 3). Washington, DC: The World Bank.

Nord, T. (2007). *Corporate Social Responsibility in the Hotel Industry: A Cross Cultural Perspective.* Master's Thesis. Stockholm: Stockholm University.

Rubin, G. D., Ryerson, C. J., Haramati, L. B., Sverzellati, N., Kanne, J. P., Raoof, S., Schluger, N. W., Volpi, A., Yim, J-J., Martin, I. B. K., Anderson, D. J., Kong, C., Altes, T., Bush, A., Desai, S. R., Goldin, J., Goo, J. M., Humbert, M., Inoue, Y., Kauczor, H.-U., Luo, F., Mazzone, P. J., Prokop, M., Remy-Jardin, M., Richeldi, L., Schaefer-Prokop, C. M., Tomiyama, N., Wells, A. U. and Leung, A. N. (2020). The role of chest imaging in patient management during the COVID-19 pandemic: A multinational consensus statement from the Fleischner Society. *Radiology*, 296(1), pp. 172–180.

Schwarz, L. A. (2018). *Villanova University's Sustainability Plan: The United Nations Sustainable Development Goals Applied to Higher Education.* Villanova: Villanova University.

Scott, D., Hall, C. M. and Stefan, G. (2012). *Tourism and Climate Change: Impacts, Adaptation and Mitigation.* Abingdon: Routledge.

Swarbrooke, J. (1999). *Sustainable Tourism Management.* Wallingford: CABI.

Welford, R., Ytterhus, B. and Eligh, J. (1999). Tourism and sustainable development: An analysis of policy and guidelines for managing provision and consumption. *Sustainable Development*, 7(4), pp. 165–177.

14 Virtual Tourism and Rebuilding the Tourism Industry in the Post-COVID-19 Era

Fernando Mayor-Vitoria

Introduction

Tourism has become almost a basic necessity in the most advanced societies. In fact, an increasingly widespread indicator of quality of life is the percentage of families who cannot afford a week's vacation every year, making tourism an inherent feature of the welfare state. Thus, the tourism industry moves billions around the world and has consolidated itself as one of the basic pillars of the gross domestic product (GDP) of the main economies. The logic of the capitalist market quite insistently invites us to make increasingly intensive use of our leisure time, inoculating us with the idea that we must make the most of that well-deserved rest after hard work. Without that rest the current economic system would not be sustained.

This rest is a right that has almost become an obligation. Leisure stimulates business. However, this tourist pressure has caused several sustainability problems for the environment, culture, and society in general. The term "overtourism," understood as the tourist overcrowding of destinations or overflowing of their reception capacities (Milano, 2018), has recently settled in the usual vocabulary of researchers, technicians, and tourism professionals as a disappointing allegory of the failure of tourism sustainability policies designed to date. It is in this context that currents of thought coming from the 1980s, such as ecological economics, steady state economics (Daly, 1992), or prosperity without growth (Klingseisen and Jackson, 2011), are beginning to come back into fashion.

In the middle of the debate on sustainability, destination saturation, or overtourism, in some tourist areas there is a standstill, a slowdown caused by COVID-19, a phenomenon unknown since tourism became an industry. In this context, there is an opportunity to rethink tourism and get out of this crisis by promoting a new tourism model that corrects the vices acquired by the frenetic growth that this activity has been experiencing (Schneider et al., 2010).

This crisis has generated a period of great uncertainty and distrust that has led us to situations which were unthinkable until very recently. It must be admitted that in the last decades, facts and circumstances have appeared that are transforming the very essence of society. Globalization, climate change, unlimited growth, the crisis induced by the international financial system, and

DOI: 10.4324/9781003315179-19

the progressive disappearance of ethical rules in the social sphere between groups and nations have led to alarming situations that manifest themselves in the most diverse areas, from growing socio-economic inequality to natural catastrophes. The crisis caused by COVID-19 has a deeper dimension, both individually and socially, as it affects not only the survival of individuals, but also society at all its levels.

In the field of tourism, the effects of the coronavirus are strongly conditioning activity, since it directly affects the bases of the tourism system: mobility and contact between people. It has been well studied that tourism activity is vulnerable (Song and Lin, 2010), and in the face of a health emergency, tourism is not a priority (Solís et al., 2012). In addition, the evolution of the pandemic at the global level makes it impossible to predict when the end of this crisis will be, because some studies on possible scenarios speak of a recurrent rebound in different months (Huertas et al., 2020). As a result, there is growing uncertainty and fear of the disappearance of a large part of tourism activity in the short term, with all that this would mean for those economies in which the tourism industry is the mainstay.

In this context, it seems complicated to venture figures and impact forecasts. The scientific community, international institutions, and even some renowned consultancy firms have set to work in the midst of the pandemic, with a proliferation of quantitative studies carried out in record time, in an attempt to estimate the numbers of the disaster. A cross-sectional reading allows people to identify another point of coincidence with those studies on other tragic catastrophes experienced, such as tsunamis, earthquakes, fires, and even terrorist attacks. However, there are substantial aspects that differentiate it. In fact, this crisis does not destroy infrastructure or physically devastate the territory. This implies that it cannot be confined to a specific area. Moreover, it is not known when it will end, which forces us to formulate different scenarios with different levels of destruction of wealth and suffering.

It is urgent to know the magnitude of the impact and, as a scientific community, we must act thinking in the future. Obviously, this will be done after the most rigorous analysis and diagnosis possible, proposing dynamic scenarios, and trying to take into account the large number of factors and variables involved in this complex situation. But in statistics there is a maxim: the greater the degree of uncertainty in the model, the less reliable the predictions, as "the only certainty is uncertainty" (Buhalis et al., 2020).

In the short history of tourism, we have never faced a phenomenon with these characteristics, so it is necessary to value collective intelligence as a key element of innovation and creation, in order to find the way out of this crisis, providing research work to help solve present and future problems. With the certainty that the socio-economic impact will be calculated with great precision when all this happens, and holding on with caution to the currently existing data and forecasts, a series of pieces of evidence are noted. On March 16, 2020, the WTTC estimated a loss of 50 million jobs in the tourism sector and on March 25 this estimate reached 75 million jobs (Jaipuria et al., 2021). In this sense, the UNWTO, on March 5, 2020, estimated a decrease of 1–3% in international tourist arrivals, in relation to the expected growth of 3–4% in January 2020 and an economic impact of between 30 and 50 billion dollars in lost spending by international visitors (Škare, 2021). Twenty days later their forecast was raised to a 20–30% reduction in tourists and US$300–450 billion, 30% less than in 2019. In early April, UNWTO stopped publishing forecasts on its website.

It is a patent fact that the current situation has literally frozen tourism activity even with the already well-known repercussions for destinations. The socio-economic drama is brutal, not only for the large companies with a certain capacity to maneuverer and hold out to save the situation, but also for those small and medium-sized hoteliers, restaurants, guides, agencies, etc., that depend on government aid and incentives. In the best-case scenario, travel could resume three months after the declaration of the pandemic and with great restrictions.

In reality, these data are as interesting as they are volatile. It is important to consider what factors will condition the recovery of tourism activity. Tourists' fear of risk and disposable income will be key elements in the rate of recovery of demand, since those people with less fear of becoming infected and with a higher level of income will begin to travel earlier. The first of these elements, the degree of risk, will diminish over time, in the sense that health security will be strengthened as the crisis is brought under control. Second, many families will see their incomes seriously reduced. However, purchasing power or capacity is relative, since it depends on market prices, and there is a high possibility that prices in the tourism sector will be significantly reduced (Hall et al., 2020). It is also to be expected that household incomes will recover over time, and here we enter into unreliable speculations, because it will depend largely on the policies of the European Union.

Analysis of Recovery Strategies

In this analysis of recovery strategies, it will be checked whether the crisis is translating into a real change of model towards a tourism system with higher levels of sustainability (Cabello et al., 2019). For this purpose, the following paragraphs are focused on the strategies being defined in the European Union and its different regions with a tourism tradition, given that the European tourism system constitutes the world's leading tourist destination with more than 716 million international tourist arrivals in 2018, representing 51% of international tourist flows and 39% of global revenue, which accounts for nearly 10% of the GDP of the European Union. Such analysis will make it possible to compare how different destinations are monitoring, supporting, communicating, and carrying out the reconstruction of a sector that has been severely affected by the restrictions imposed on travel by the World Health Organization. This process is completed with a comparison of current measures with those implemented during the 2003 SARS health crisis in Asia.

Analysis of Strategies Proposed by the European Commission

The European Union's response to the COVID-19 pandemic can be found in the Commission's communication to the Parliament, the Council, the Economic and Social Committee, and the Committee of the Regions "Tourism and Transport in 2020 and beyond." It sets out the common European roadmap for the lifting of the coronavirus containment measures and includes a package of measures for the recovering of the transport sector in order to ensure that the Union retains its position as the world's leading tourism destination.

Among the immediate measures to alleviate the situation, the Union has launched initiatives to solve the problem of the lack of liquidity of many players in the sector, unlocking financial funds and recommending the offer of vouchers to passengers and travellers as an alternative to travel reimbursement; it has provided funds for employment protection with an emphasis on improving the skills of workers and the unemployed to cope with a green and digital transition.

Finally, it is committed to the promotion of local tourism, i.e. national and inland tourism in the Union, a modality that will be reinforced with the proposal for the European Year of Railways 2021 and which is the Commission's strategic approach to a sustainable blue economy that will drive the transition to a more sustainable model of coastal and maritime tourism, also supporting agrotourism as a sustainable model in rural areas.

In addition to the immediate measures, the Commission's ambition with regard to tourism is to maintain Europe as the world's leading tourist destination in terms of value, quality, sustainability, and innovation, and this vision will be supported with dedicated financial instruments and investments at the different territorial levels. The central element of the new strategy to address the longer-term common challenges is sustainability contributing to the European Green Pact, the objectives being to achieve affordable, sustainable transport, improve connectivity, boost smart management of tourism flows, diversify the tourism offer by fighting seasonality, highlight cultural and landscape diversity across Europe, improve workers' training, especially sustainability-related skills, and promote sustainable accommodation through the adoption of eco-labels, management systems, and environmental auditing.

The second pillar would be the digital transition as a potential element to optimize the management of flows, to be able to offer a more efficient use of resources, and to better understand the needs of tourists and their trends. Firstly, the Commission's communication does not identify a discourse that can be directly connected to post-growth tourism strategies, as it does not question the current tourism model nor does it refer to the conflicts arising from a possible excess of tourism in destinations or its possible manifestations such as congestion, privatization of public spaces, overloading of infrastructure, loss of a sense of belonging to the place, or the dismantling of socio-cultural connectivity pointed out in the studies by Milano (2018). Although it is true that some of the listed long-term objectives could be considered as attempts to mitigate the effects of tourism such as, for example, depersonalization and diversification, until the European Agenda for Tourism 2050 is finalized, it cannot be claimed that the new strategy will impose limits to growth and transcend the prevailing tourism models.

It should be noted that the Commission's measures are no more than recommendations, while the final decision depends on the member states, a reason that justifies the analysis of the responses of the member states through the strategies and actions of their tourism regions. To sum up, a selection of case studies has been analysed that have in common their participation in Network of European Regions for Competitive and Sustainable Tourism (NECSTOUR), a European network of 34 tourism regions and 30 associate members (universities, research institutes, sustainable tourism associations, and representatives of tourism companies) which works proactively sharing best practices and establishing alliances so that destinations can achieve a smarter and more sustainable approach to tourism development.

Strategies of the Regions Belonging to NECSTOUR

The actions implemented against the pandemic by the following regions have been reviewed: Belgium (Flanders, Wallonia, and Brussels), France (Île de France and Occitania), Italy (Liguria, Sardinia, and Tuscany), the Netherlands (Limburg), Spain (Catalonia, Andalusia, Balearic Islands, Canary Islands, and Valencia), Croatia, Ireland (Donegal), Denmark (West Coast), and Scotland. It is a sample of 18 destinations distributed throughout Europe that receive millions of tourists every year. Among them are urban destinations, cultural tourism destinations, coastal destinations associated with maritime tourism, and the sun and beach seasons. The measures taken by the territories that make up the study area are grouped into several categories: information and monitoring, governance, tax relief, liquidity, worker protection, technical assistance, data analysis, promotion, recovery plan, and tourists.

The first group of measures, information and monitoring, refers to the initiatives that the destinations have put in place to report on the legislative measures, aid, regulations, and protocols that governments have adopted in the wake of the coronavirus. Twelve of NECSTOUR's member regions have set up dedicated websites and toll-free telephone lines to provide information about COVID-19 to tourists and entrepreneurs.

The governance category refers to collaboration between public administrations or the creation of crisis management teams, so that in ten of the cases analysed some of these measures were taken in order to provide a coordinated and effective response. In this section, the creation of a crisis management team in Tuscany, which integrates the regional government in the field of crisis management, stands out; the regional government for tourism, the regional digital agency, the municipalities, the airport and port, among other key players. Limburg should also be noted for creating a regional force that is made up of both the regional government and business and private sector representatives.

Public-private collaboration also occurs in the case of the Valencian Community, characterized by intergovernmental coordination, as different regional agencies have worked closely together to develop action protocols and extraordinary measures such as, for example, medicalized hotels.

In terms of protection for workers in the sector, the contributions of self-employed workers have been reduced or cancelled, and unemployment programmes have been made more flexible to cover all those who have been unable to work due to the confinement in Spain, Italy, and Belgium. Among the technical assistance measures, some regions such as Occitania, Andalusia, Valencia, Catalonia, Tuscany, and Scotland have worked to ensure distance learning for the unemployed, either through learning platforms or through digital workshops or thematic webinars.

Actions related to data analysis have also been implemented as periodic surveys in Valona to measure the impact of the coronavirus on the sector. For the same purpose, in Occitania, weekly monitoring is carried out with the business federations. In Tuscany, periodic studies of the issuing markets have been offered. Scotland, Andalusia, the Canary Islands, and Catalonia have been the most proactive in this area, developing a variety of tools to provide data to guide actions to combat COVID-19: a weekly survey of managers of regulated tourist accommodation, sentiment analysis, and active listening on social media and the Internet, as well as the preparation of weekly reports with data on flight searches, bookings, and seat capacity offered by airlines.

Another set of actions corresponds to destination promotion campaigns consisting of audio-visual productions aimed at attracting potential tourists. In this case it can be noted that almost all regions have carried out some digital advertising campaign during the confinement encouraging tourists to stay at home, reminding them of their tourist attractions, and encouraging them to travel when normality returns. It should be noted that the Spanish regions of the Balearic Islands, Catalonia, and Andalusia are working on a medium- and long-term tourism recovery plan, but there is still no public information to assess whether they include post-growth strategies.

Lastly, there are actions aimed at tourists, including vouchers, which seek to ensure that tourists affected by COVID-19 do not cancel their trips and postpone them, as has happened in Belgium and the Netherlands, measures that seek to stimulate tourism demand in certain segments. In the case of Spain, destinations such as Galicia and Malaga offer special conditions or tourist vouchers to use the public healthcare services. In Italy they have gone further, and low-income families will receive a tourist voucher of 500 euros to spend in hotels and hostels in the country.

Analysis of the World Tourism Organization Recommendations

The actions of the European regions are in line with the priorities for tourism recovery set by the World Tourism Organization (UNWTO), which can be summarized as: liquidity and job protection, recovery of confidence based on security, public-private collaboration, the opening of borders with responsibility, harmonizing protocols and procedures, digital transformation of jobs, and innovation and sustainability as pillars.

However, sustainability and innovation are not exactly new values that have burst onto the tourism scene as a result of the health crisis, and, moreover, they are always framed within a logic that does not renounce the growth of the activity, while trying to make it compatible with respect for the environment.

There are some signs of how these types of measures can take shape or become a reality, advocate digital transformation and the consolidation of innovation ecosystems, innovation in business models, acceleration of the transition to a circular economy, reducing the carbon footprint to mitigate the climate change crisis, defining load capacities, and advancing in the effective management of destinations based on systems of indicators and monitoring systems, as the organization itself is already doing through the International Network of Tourism Observatories (INTO).

Of all the priorities, the one that could have the greatest impact on the creation and promotion of a new tourism development paradigm is the one that proposes innovation and sustainability as the new normal, which, according to the UNWTO, requires a new approach and a new logic of systems and networks. In reality, it once again comes up against the casuistry presented above in the European Union, since the UNWTO is an international organization attached to the United Nations that promotes tourism, but has no competence to impose its priorities and recommendations on member states.

Virtual Experiences

World Heritage Sites are considered heritage properties of great value by all on the basis of their recognized outstanding universal value. Tourism in its social dimension and compatibility with protection and conservation has been one of the United Nations Educational, Scientific and Cultural Organization (UNESCO)'s proposals for these properties for a decade (Gravari-Barbas, 2015). Thus, both information and interpretation are part of the tourist experience in which an essential role is played by guided tours in this symbiosis between heritage and tourism.

In recent years, society has been increasingly exposed to virtual reality environments in the media, and high expectations about this virtual environment have been developing. This use of new technologies is also being integrated into the dissemination of heritage and, in turn, into tourism activities.

This trend has been accentuated with the arrival of COVID-19 which has given a new meaning to virtual visits as the only way to maintain a heritage and tourist activity. Virtualization will mean that, at a time of crisis, when monuments and museums are being closed, visits can continue to take place virtually. With all this, the arrival of this pandemic, in addition to bringing about the expected changes due to the alteration of the tourist activity itself, is introducing innovative activities through virtualization that affect heritage centres, destinations, and even the behaviour of the public.

Innovation in the field of cultural heritage is presented as the alternative capable of providing added value to the visitor, through technological innovations, new organizational processes, new modes of exhibition, innovation in communication, new relationships with the same audiences, and new ways of communicating with the public (Camarero et al., 2015). Within the field of innovation, new technologies and cultural heritage are one of the major fields of study.

While in tourism new technologies and their study have focused on training, hospitality, employment, travel agencies, product creation, demand analysis, and marketing (Huang et al., 2016), in the cultural heritage sector there is an intense scientific production focused on education, conservation, the application of technologies to the territory, geolocation, and the application of virtual reality (Elordi et al., 2012).

Innovations in the way of transmitting and communicating thanks to new technologies, at a time of increased travel and the appearance of new ways of traveling such as cultural tourism. But these practices have also allowed a rapid reaction and adaptation to the new circumstances in the face of the COVID-19 crisis that has come to mean the end of mass visits.

The new scenario caused by the pandemic has forced rapid interventions and proposals from institutions and research centres that have reacted quickly to the new circumstances with working groups at the institutional level. In museums and heritage centres there has also been a rapid reaction to the COVID-19 crisis. Thus, together with technological practices such as thematic videos, concerts, educational resources, apps, and their own web pages that were already in place, some museums and heritage centres have also reacted quickly to the COVID-19 crisis. Some centres launched proposals such as online courses (Museo Thyssen, Museo del Prado), videoconferences (British Museum), presentation of works (Atelier des Lumières, Paris), virtual tours (Castilla y León, Patrimonio de la Humanidad).

Another alternative was the initiative to organize virtual tours with guides. More precisely, the Uffizzi Gallery organized virtual tours with guides in different European cities. In all cultural heritage, visits and tours are the basis, whether face-to-face or virtual, and play a role in the visitor's experience by providing interpretation and information through the figure of the guide. Studies are known on the construction of adequate speeches in the interaction of the guide with the tourist, taking into account the tourist and their stories as an image of the place and a source of information (Cunin and Rinaudo, 2008).

Recommendations and Conclusion

The analysis of the proposed actions is mostly aimed at returning as soon as possible to the expected growth path, which was abruptly interrupted in March 2020. To date, the recovery strategies are mostly defined as short-term measures, without considering a longer time horizon.

In this sense, there are few initiatives that openly defend taking advantage of this crisis, and the unforeseeable implications it will have for the perception of demand (mainly to avoid overcrowding), to seek an alternative method of destination management that prioritizes respect for reception capacities and flow management over the massive promotion of destinations and price and product strategies to capture more demand than their competitors at whatever cost is necessary. Few destinations have tried to reach a consensus among their main stakeholders on recovery strategies by looking at the mistakes of the past in order to try to avoid them. Few tourism stakeholders have seen this downturn as an opportunity to improve the quality and sustainability of destinations.

The rapid changes in the environment and the new scenario posed by the health crisis caused by COVID-19 have caused us to review the way we live, work, and travel. These new circumstances, together with the change in visitor behaviour, as well as technological advances, have made it possible for cultural institutions to rethink their strategies and their forms of management, with many of them focusing on innovation to develop new ways of interpreting heritage.

These new ways can be beneficial in attracting new audiences through the application of technologies in their facilities and in their communication strategies. Undoubtedly, the new post-COVID-19 environment demands that the global tourism industry reboot itself, both in the way we communicate and in the way we travel and get to know new places.

As discussed in this chapter, the transformation from traditional tours to virtual tours has achieved a greater reach in terms of the public, not only because of the increase in visits but also because it has reached audiences that had not been reached until then.

Bibliography

Buhalis, D., López, E. P. and Martinez-Gonzalez, J. A. (2020). Influence of young consumers' external and internal variables on their e-loyalty to tourism sites. *Journal of Destination Marketing and Management*, 15, p. 100409.

Cabello, J. M., Navarro-Jurado, E., Rodríguez, B., Thiel-Ellul, D. and Ruiz, F. (2019). Dual weak–strong sustainability synthetic indicators using a double reference point scheme: the case of Andalucía, Spain. *Operational Research*, 19(3), pp. 757–782.

Camarero, L. A., Álvarez, S. V. and Pérez, F. G. (2015). Offshoring de producción intermedia e innovación:¿ Afecta el modo de gobierno a la relación?. *Economía Industrial*, 397, pp. 105–119.

Chien, P. M. and Ritchie, B. W. (2018). Understanding intergroup conflicts in tourism. *Annals of Tourism Research*, 72, pp. 177–179.

Cunin, E. and Rinaudo, C. (2008). Consuming the city in passing: guided visits and the marketing of difference in Cartagena de Indias, Colombia. *Tourist Studies*, 8(2), pp. 267–286.

Daly, H. E. (1992). Steady-state economics: concepts, questions, policies. *Gaia-Ecological Perspectives for Science and Society*, 1(6), pp. 333–338.

Elordi, U., Segura, A., Goenetxea, J., Moreno, A. and Arambarri, J. (2012). Virtual reality interfaces applied to web-based 3D E-commerce. In *Engineering Systems Design and Analysis*. American Society of Mechanical Engineers, pp. 341–350.

Gravari-Barbas, M. (2015). Winescapes: tourisme et artialisation, entre le local et le global. *Cultur-Revista de Cultura e Turismo*, 8(3), pp. 238–255.

Hall, M. C., Prayag, G., Fieger, P. and Dyason, D. (2020). Beyond panic buying: consumption displacement and COVID-19. *Journal of Service Management*, 32(1), pp. 113–128.

Huang, Y. C., Backman, K. F., Backman, S. J. and Chang, L. L. (2016). Exploring the implications of virtual reality technology in tourism marketing: an integrated research framework. *International Journal of Tourism Research*, 18(2), pp. 116–128.

Huertas, A., Oliveira, A. and Girotto, M. (2020). Gestión comunicativa de crisis de las oficinas nacionales de turismo de España e Italia ante la COVID-19//Crisis communication management by the national tourist organizations of Spain and Italy in the face of COVID-19. *Profesional de la información*, 29(4), pp. 4–18.

Jaipuria, S., Parida, R. and Ray, P. (2021). The impact of COVID-19 on tourism sector in India. *Tourism Recreation Research*, 46(2), pp. 245–260.

Klingseisen, A. and Jackson, A. P. (2011). Mechanisms and pathways of growth failure in primordial dwarfism. *Genes and Development*, 25(19), pp. 2011–2024.

Milano, C. (2018). Overtourism, malestar social y turismofobia. Un debate controvertido. *PASOS. Revista de Turismo y Patrimonio Cultural*, 16(3), pp. 551–564.

Ortega, G., Navarro, E., Cerezo, A. and Torres, E. (2020). *Turismo poscoronavirus, ¿una oportunidad para el poscrecimiento. Turismo pos-COVID-19: Reflexiones, Retos y Oportunidades; Cátedra de Turismo CajaCanarias-Ashotel de la.* Canarias: Universidad de La Laguna, pp. 161–173.

Schneider, S., Ricci, F., Venturini, A. and Not, E. (2010). Usability guidelines for WAP-based travel planning tools. In U. Gretzel, R. Law and M. Fuchs (Eds.), *Information and Communication Technologies in Tourism 2010*. Vienna: Springer, pp. 125–136.

Škare, M., Soriano, D. R. and Porada-Rochoń, M. (2021). Impact of COVID-19 on the travel and tourism industry. *Technological Forecasting and Social Change*, 163, p. 120469.

Solís, M. H., Martínez, A. M. and Oromendía, A. R. (2012). Impact of the international financial crisis on the Spanish tourism sector. *GSTF Business Review (GBR)*, 1(4), p. 149.

Song, H. and Lin, S. (2010). Impacts of the financial and economic crisis on tourism in Asia. *Journal of Travel Research*, 49(1), pp. 16–30.

15 Technology and Safety of Aviation and Tourism

The Past, Current, and Post-COVID-19 Trends

Azlina Mohd Ali, Nor Aida Abdul Rahman, and Azizul Hassan

Introduction

Tourism and aviation complement each other since tourism activities depend on the available transportation to move the traveller from one location to another location. Tourism and aviation are acknowledged as an economic sector that contributes to world gross domestic product (GDP) for many countries and regions, both in developing and developed countries. According to Koo, Park, and Lee (2017), the tourism sector is also interconnected with other fields such as geography, economics, and sociology. The safety and security of travellers throughout their journeys are the key concern to support the growth of the tourism industry. The safety of the travellers is often the main focus for travellers; however, it is rarely being discussed in previous literature (Santana-Gallego et al., 2019; Koc, 2016).

On the other hand, aviation is recognized as a key transportation mode that supports both local and international tourism activities and tourism markets. In the aviation context, any developments in the aviation industry will significantly contribute to the growth of the tourism industry. Aviation has become the key component that enables travellers to move from one destination to another destination. Globally, the world airlines market is expected to reach US$744 billion by 2026, and according to the World Tourism Organization, the tourism industry will continue to grow even after being hard hit by the COVID-19 pandemic. In spite of the multiple changes, new norms, and shocks from man-made crises, the pandemic, economic crises, natural disasters, and others, both the aviation industry and tourism will continue to grow. Despite all of this, issues with regard to the safety and security of travellers are the key essential to focus on. The safety of travellers during their travel and tourism activities is essential to ensure these two industries grow successfully in the future.

As mentioned earlier, tourism and aviation complement each other since tourism activities depend on the available transportation to move the traveller from one location to another location. Tourism and aviation are acknowledged as an economic sector that contributes to world GDP for many countries and regions, both in developing and developed countries. In the 21st century,

DOI: 10.4324/9781003315179-20

technology brings enormous opportunities for both tourism and the aviation sector to bring innovation in their operation and business activities, to improve their systems, and to help travellers to travel safely during their journey (Rahman et al., 2019). For example, from aviation and tourism perspectives, what are the technologies available in order to keep travellers safe and protect them from any harm or dangerous risks?

Technology has been identified as one of the key components in helping the safety and security system in both industries. Technology is critical in the tourism and travel sectors to secure the travellers' safety throughout their journey. As highlighted by Burmaoglu, Sartenaer, and Porter (2019), technology has been regarded as a contributor to economic development for 50 years. At present, with the development of many new technologies in this fourth industrial revolution, there is a significant need to investigate the role of technology in every type of industry including tourism in ensuring the safety of travellers during their tourism activity. Improved safety initiatives should help disperse visiting spending to a wider spectrum of tourist facilities, such as restaurants, theatres, and leisure facilities, and away from hotels and tourist resorts (George 2003). In general, the tourism and aviation industry is always looking for a solution to improve its business efficiency and operation, and to ensure the safety of its passengers or travellers. The aviation sector always supports tourism and other business activities including the movement of cargo worldwide. Current aircraft technology is designed for improvement in fuel burn compared to similar aircraft of a decade ago. This new technology-driven aircraft has resulted in lower emissions and better environmental sustainability. Additionally, improved aircraft performance may result in improved airworthy of the aircraft and travellers.

Having this as a background, the arrangement of this chapter is related to three main aims of this study: firstly, technology in the aviation and tourism industry. Secondly, to understand how technology could ensure the safety and security of travellers throughout their journey by highlighting the key technology used in aviation and tourism. Thirdly, to provide discussion and review from past, current, and future study in relation to safety in the field of aviation and tourism. This chapter is then closed with a summary of safety in the aviation and tourism context, highlighting future research opportunities for future researchers. Prior to highlighting these three aims developed in this study, the authors will first discuss the notion of technology and safety in the area of tourism and aviation.

Safety Definition: Safety Perspective from the Aviation and Tourism Sector

Safety is a popularic term meaning the freedom from any danger or risk, and also relates to any activity or effort to protect something from being harmed or exposed to any injury or danger. It is a popular term used in many industries such as in construction, healthcare, manufacturing, maintenance, services, and the transportation industry including the aviation and tourism industry. Safety is one of the key concerns in any operation that poses a high risk of job injury.

Safety also refers to any activities or operations that pose a high risk of injury to people or travellers, or that expose people or travellers to hazardous materials or resources, and all operations entail some degree of risk (Kraus et al., 2018). Safety in the context of aviation refers to the condition in which the possibility of harm to persons or of property damage is reduced to, and preserved at or below, an appropriate level by a continuing process of hazard identification and safety risk management (International Civil Aviation Organization, 2018).

In aviation, the concept of safety and safety management has been important since 1903 and the Wright brothers' first flight. Safety from a flight perspective refers to the safety of the passengers who board the aircraft. The airworthiness issue is the top priority in the aviation sector (Musa and Isha, 2021). The complexity of the cargo operation and other aviation sectors such as airline catering and other ground handling service including maintenance, repair, and overhaul (MRO) requires comprehensive safety management system (SMS) implementation. While safety in the tourism sector refers to the threat to travellers connected to crime rates, terrorism, food safety, health issues, and natural disasters. This topic gained popularity in travel and tourism research after the 9/11 tragedy, and since then academicians and practitioners have started seeking the best solutions to mitigate these negative impacts.

For the Malaysian aviation and tourism industry, the image as a safe and secure destination was tarnished a few years ago, and new unfortunate incidents (such as those of MH 370 and MH 17 in 2014) have brought more damage to Malaysia's image. Health issues, terrorism, Lahad Datu intrusion, repeated kidnappings and shootings in Sabah, the twin airline incidents of MH 370 and MH 17, riots and illegal demonstrations, and false reporting by international media bring new challenges to Malaysia. Although some incidents may have had a short-term impact on Malaysia's tourism industry, they still have a big impact on Malaysia's branding process. Many travellers and Malaysians themselves still believe that Malaysia is a safe destination and country to visit and stay in, but more outstanding efforts were required to make sure that Malaysia's tourism industry was able to recover from this negative impact as quickly as possible.

Therefore, safety is the top priority issue in travel and tourism activities. The application of technology in aviation and tourism activities may increase the level of safety for travellers or passengers. With current technology adoption, the threat of possible harm could be reduced and preserved at a very minimal level. In other words, safety refers to the condition of being protected from any harm, injury, or danger.

The development of aviation safety is continuously progressing throughout the years. Essentially, the evolution of safety in the aviation sector could be divided into three main eras, namely: (1) technical, (2) human factors, and (3) organizational era, which started in the 1950s and continues today; see Figure 15.1.

In the aviation sector, there is always a threat to the safety of the aircraft and the passenger; thus every member in the aviation sector is responsible for cultivating safety culture in the workplace. Accordingly, in the earlier study by Gill and Shergill (2004), they mention that the effectiveness of the safety management

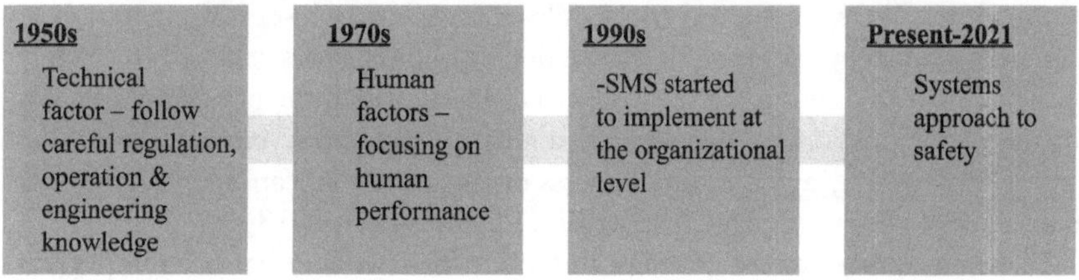

Figure 15.1 The technical, human factor, organizational, and technological eras of safety in the aviation sector (source: developed by the authors).

system is essential to manage to any threats. In fact, organizational practices are acknowledged as the key factor for any reliable safety system and cultivating safety culture is a must in any organization (Reason, 1997). It is important to establish safety culture at the workplace in any organization, especially in a transportation field such as aviation. With a safety management system, there is a reactive approach, resulting in declining rates of accidents until the mid-1980s. As a result, the fatality rate in the air transport sector remained stable even though during that time traffic was increasing. It is acknowledged that the connection between the accident rate and technology adoption is critical as technology application could have a potential effect in preventing loss of life and injuries and reduce fatalities (Ali et al., 2019).

Much past research, however, highlighted the question of whether there is any relationship between technology adoption and the safety of air travellers for instance. As far as the aviation industry is concerned, technology application in many parts of ground aviation activities is being increasingly introduced to organizations to ensure the safety of the travellers or passengers. For instance, self-check-in technology, baggage drop technology, and baggage screening technology are all related to the safety of travellers. It is recognized that safety management includes systematic activities and comprehensive effort in managing accidents, and it is vital to have safety culture practiced in the travel and tourism organization at all levels of staff. The psychological aspect of safety when embedding safety into the organizational culture in any organization could lead to high awareness among the travel and tourism staff to effectively be a team member in the organization, which could expedite the development of a good culture practice (Roussin, 2016). For instance, in the United Kingdom, safety management activities in the aviation sector, which include flight operation, engineering, and ground handling organization, has achieved high levels of safety performance due to increased levels of safety culture among the MROs staff. In fact, in the Eastern countries including Malaysia, a comprehensive process for managing risk across the whole aviation sector is also being discussed and improved. However, having reviewed the literature, again, there are not many studies that focus on the aviation sector especially in relation to travel and tourism organizations and technology adoption. Again, this shows a significant gap that needs to be filled by the current study.

Safety and the Tourism Sector

As highlighted earlier, in the tourism sector, safety and security are connected with providing quality. More than any other economic activity, the success or failure of a tourism destination depends on being able to provide a safe and secure environment for visitors. This was highly evident in the aftermath of the tragic events of 11 September 2001. In the week following the 11 September attacks in New York and Washington, passengers abandoned airports in large numbers as the effects of these attacks extended beyond US border. After that event, many issues arose with regard to the safety of travellers across the globe. Starting from that point, security in tourism activities in every country has been stricter than before. Biometric technology and other technology applications have been introduced to ensure the safety of travellers.

As highlighted earlier, in the tourism sector, safety from the travellers' perspective means that travellers feel safe to travel throughout the journey without having a negative impression of the destination they visit (George, 2003). The feeling of not being safe during travel may give a negative impression of the destination.

Technology and the Safety of Travellers

Technology is always related to the safety of travellers. For instance, technology such as the Internet of things, mobile technology, drones, robotics, and many others could help both aviation and the tourism sector.

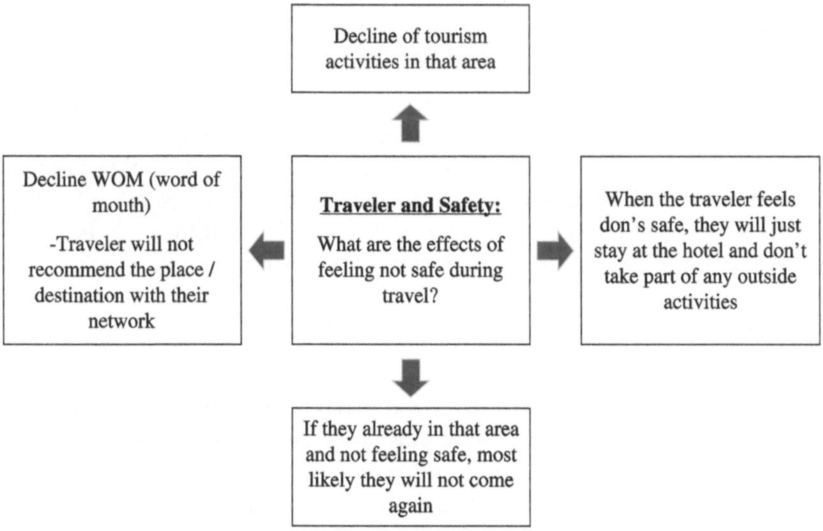

Figure 15.2 What happens if travellers do not feel safe during travel (source: developed by the authors).

Mobile Technology and Safety in Travel and Tourism: Before and after the COVID-19 Era

Despite the steady travel activities across the globe, issues with regard to the safety of travellers remain the most significant issue highlighted by many scholars (Figure 15.2). Safety has been highlighted as a central issue in the era of the COVID-19 pandemic, as well as before the COVID-19 outbreak. Among the issues of safety related to tourism is crime in specific areas, also known as tourist victimization. Having technology in hand during travel may reduce the possibility of being a victim during travel and tourism activity. At present, technological applications in tourism such as mobile technology are very helpful in increasing the safety of travellers during travel. Apart from using mobile technology as a main reference for tourist guiding, it may also reduce the chances of being a victim of any crime at the visited areas or destinations. Mobile technology may help travellers in many ways. As highlighted by previous study, mobile technology could help travellers browse and book their flight or event tickets without having to visit the premises. Mobile technology helps travellers to save time and increase their safety.

Technology can also deliver a seamless travel experience and reduce the risk of exposure to any disease such as COVID-19. Additionally, mobile technology also helps travellers in providing destination guides and information on currency conversion, as well as helping travellers with itinerary creators. Mobile technology also enables travellers to download language apps which help travellers in communication throughout their journey. Mobile technology benefits travellers in many forms. It not only provides the traveller with useful information, but also gives good entertainment while travellers are away. Travellers can use technology for reading, communicating, guiding, browsing, and streaming music and videos.

Automation Technology in Aviation and Tourism

In efforts to encourage compliance with corporate policies, and to ensure the safety of travellers, automation is another example of technology that could be useful. Similarly, when individual travellers feel unsafe, they'll be able to push a button to be directed to a safer location – efficiently and without the friction of having to wait on hold on the phone. The travel management system will automatically generate an itinerary to bring the traveller home within policy.

Touchless Interface Technology in Aviation and Tourism

Another development in the travel industry is contactless travel. For example, cars connected via the Internet of things (IoT) can be opened electronically with a smartphone – no key needed. Along those same lines, travellers are be able to use their smartphone to open their hotel room door, without even turning the knob. Many hotels already use near-field communication (NFC) technology for this purpose. On the other hand, the hotel's technology then geolocates you, recognizes you, and confirms your credit card is on file. Then the technology sends an update to the mobile app with your room number. You go to your room, hold the phone near the door, and the door opens. This touchless interfacing technology is likely to extend to flight ticketing, identification verification, check-ins, purchasing, ordering food, and more. Again, this type of technology could minimize the negative threats to travellers and keep travellers safe throughout their journey.

Voice Search and Voice Control Technology

Another interesting recent technology in aviation and tourism is known as voice search and choice control technology. This technology, with the help of smartphones, smart speakers, and AI assistant devices, has helped to increase the relevance of voice search as far as technology trends in the travel industry are concerned. In particular, a growing number of travel customers are using voice search to find and book airline tickets, hotel rooms, and travel experiences. On top of this, voice control is also playing a growing role in actual travel experiences too. Within hotel rooms, voice-controlled devices can be used to control lighting and heating, or to obtain tourist information without needing to speak to a member of staff. This is likely to continue to expand into other areas of the industry too. Again this technology helps travellers to ensure their safety throughout their travel journey.

Conclusion

As a conclusion, it is time for governments to reconsider the role of technology in the aviation and tourism sector to help increase the safety of travellers whether in a local or international context. People or travellers on the move should be able to own their digital biometric profile and have the ability to push this secure data in advance to make their journey easier. Traveller participation will enable the wider use of pre-clearance and will make international border crossings more efficient. Having this as background, therefore, it is vital to have the entire process of border management used by most travellers be a wholly automated, electronic platform, built on verified biometric data technology. Again, this is significant in making sure travellers can move from one location to another location with safety, and making sure every tourism activity done by travellers is smooth and safe. They facilitate border crossings for many travellers, reduce paperwork, and allow public safety officials to direct more attention and resources to threat identification. This not only saves travellers time, but also ensures the productivity of the agency that controls the country's borders. With that, technology not only helps to ensure the safety of travellers, but also increases the productivity of the organization.

Previous tourism study has not investigated technology issues with regard to safety during travel and the use of air transport. This study could be a main platform to explore further issues on technology application in the field of aviation and tourism. In fact, this study would appear as the main reference for future scholars to investigate further issues of technology application with regard to the safety of the traveller. Essentially, this chapter has given rise to several areas for future research opportunities specifically related to travel and tourism, including transportation and safety. The tourism industry needs to provide guidelines for travellers to ensure they follow safety procedures to ensure safe travelling, etc. Further research is called for to investigate this issue in a deeper manner and enhance current knowledge and theoretical understanding of this issue.

References

Ali, A. M., Rahman, N. A. A., Mohamad, B. and Ahmad, M. F. (2019). Identifying safety management system (SMS) components in aviation industry: Key initial findings from aviation experts. *Journal of Advance Research in Dynamic and Control Systems*, 11(S1-5), pp. 1404–1408.

Burmaoglu, S., Sartenaer, O. and Porter, A. (2019). Conceptual definition of technology emergence: A long journey from philosophy of science to science policy. *Technology in Society*, 59, p. 101126.

George, R. (2003). Tourist's perceptions of safety and security while visiting Cape Town. *Tourism Management*, 24(5), pp. 575–585.

Gill, G. K. and Shergill, G. S. (2004). Perceptions of safety management and safety culture in the aviation industry in New Zealand. *Journal of Air Transport Management*, 10(4), pp. 231–237.

International Civil Aviation Organization (ICAO). (2018). *Safety Management System*. Retrieved from: https://www.icao.int/MID/Documents/2018/Aerodrome%20SMS%20Workshop/M0-2-SMS%20Overview.pdf (accessed: 20th September, 2021).

Koc, E. (2016). Risk and safety management in the leisure, events, tourism and sports industries. *Tourism Management*, 54, pp. 296–297.

Koo, C., Park, J. and Lee, J.-N. (2017). Smart tourism: Traveler, business, and organizational perspectives. *Information and Management*, 54(6), pp. 683–686.

Kraus, J., Lališ, A., Plos, V., Vittek, P. and Stojić, S. (2018). Utilizing ontologies and structural conceptual models for safety data management in aviation maintenance, repair and overhaul organizations. *Transportation Research Procedia*, 35, pp. 35–43.

Musa, M. and Isha, A. S. N. (2021). Holistic view of safety culture in aircraft ground handling: Integrating qualitative and quantitative methods with data triangulation. *Journal of Air Transport Management*, 92, p. 102019.

Rahman, N. A. A. (2019). Prominent and innovative tourism events in Peninsular Malaysia. In A. Hassan and A. Sharma (Eds.), *Tourism Events in Asia: Marketing and Development*. Abingdon: Routledge, pp. 123–131.

Reason, J. (1997). *Managing the Risks of Organizational Accidents*. Aldershot: Ashgate Publishing.

Roussin, C. J., MacLean, T. L. and Rudolph, J. W. (2016). The safety in unsafe teams. *Journal of Management*, 42(6), pp. 1409–1433.

Santana-Gallego, M., Fourie, J. and Rosselló, J. (2019). The effect of safety and security issues on international tourism. *Annals of Tourism Research*, 80, pp. 1–4. doi: 10.1016/j.annals.2019.02.004

16 The Post-COVID-19 Implications for Researching Digital Transformation in Aviation, Tourism, and Hospitality in Southeast Asian Events

Nor Aida Abdul Rahman, Azizul Hassan, Khaeruddin Sudharmin, Zawiah Abdul Majid, and Nurhayati Mohd Nur

Introduction to Aviation and Tourism

Undoubtedly the aviation industry has become more competitive and sophisticated than ever before. Following the newer forms of both marketing and management approaches, some airline organizations are gradually advancing more than others in terms of facilitating the passenger and improved services and hospitality, as well as advancing in terms of technology adoption in their organization in improving both operational and managerial performance. Many forms of technological interventions and effective engagement of the experts and brokers have made the global aviation market place more complex and intelligent. Many airliners have proven tactics for offering the most updated information to the tourists. Offering valid information also results in improved service quality. The wider availability of communication devices and particularly the introduction of the Internet have made the aviation industry worldwide with more accessible and improved connectivity within the subsectors such as airlines, airports, cargo terminals, and many more. In general, travel by air across the region has been expanding year by year. At present, travellers or passengers are offered more freedom that allows them to self-check-in and drop baggage during travel at the airport. The improved process of facilitating travellers at the airport with the help of technology adoption enables passengers to avail of a better experience. Smart technology usage by the aviation players and smartphone device technology used by tourists or travellers have improved their journey experience and made it more systematic and well planned.

Tourism events are becoming more significant in producing an incredible impact in the travel and tourism sector worldwide. Tourism events influence a wide spectrum of human activities such as politics, socio-cultural activity, and others. Globally, the market value for event tourism was valued at US$1.135 trillion in 2019 and is expected to increase to US$1.553 trillion by the year 2028 (Applied Market Research, 2021). Tourism events refer to tourism activities that

DOI: 10.4324/9781003315179-21

involve travellers moving from one destination to another destination to join a specific event such as a meeting, conference, exhibition, convention, seminar, or any other specific business reason (Van Der Wagen and Carlos, 2005). Tourism events generally involve the activities of people management, setting up the programme, logistics arrangement, food or catering management, and so on. In principle, events are organized in certain states or countries to attract more tourists to visit their country, especially in off-peak seasons.

In principle, events are regarded as a vital factor that encourages tourism activities to happen. Numerous travellers or tourists were attracted by innumerable events across the globe. Some places or destinations are visited by tourists for special events. Logically, nobody is attracted to visit a distant island or city unless some appealing events are organized or arranged at that place. Tourism events are closely related with place branding, where the event is intended to create a positive image and reputation for the place. Rahman (2019) agrees that people are travelling for events not only to attend seminars, meetings, and conferences, but also to attend certain festivals organized in certain seasons in the country. For instance, tourists love to visit Malaysia (Terengganu) for squid jigging activities, especially in May to August. As mentioned, like tourism events, people are not only travelling for meetings, conventions, and specific seasons, but also to attend specific festivals and cultural programmes including specific sports events (Olympics) and specific entertainment programmes such as theatres and concerts, as well as outdoor activities. Quinn (2010) refers to an event as an activity that brings together the specific communities.

Attraction in tourism refers to the place of interest where the tourist loves to visit. It may be connected to the events organized at that place, or because of other characteristics such as the cultural values, historical buildings. Essentially, it refers to the sector that is comprised of manmade and natural attractions which are developed to satisfy visitors' educational, recreational needs.

Previous Research in Tourism Events and Technology Advancement

The earliest study in the area of tourism events was published as early as in the 1980s. Getz (2008) in his article proposes a framework for planning event tourism. In the year 2011, Tkaczynski and Rundle-Thiele (2011) published a study on tourism events segmentation. They found that event segmentation and identifying the correct attendees for an event could be achieved via an appropriate marketing mix. The study of Jin, Qu, and Bao (2019) looked at the impact of tourism events from 2005 to 2017, focusing on the tourist flow from China to South Korea and Japan. The findings suggest that the success of tourism events depends on many factors such as political crisis, endemic disease, and financial crisis as well as natural disaster.

At present, progressive technological advancement has led to new ways of organizing events and tourism event activities. Technology has great potential to innovate business operations and improve multi industry experience as well as improve service innovation practice. In the industry, it is well recognized that technology such as robotics, as well as artificial intelligence (AI) technology, could replace expensive human labour. This technological labour is considered to be more effective and efficient in operation and goal management as it may reduce the human error cases. In the tourism context, technology advancement has allowed the tourist to stay connected with their family members, colleagues, and their work activity while they are travelling.

The recent pandemic outbreak has changed the whole landscape of travel and tourism activities globally. Touchless technology is becoming important in travelling and at the workplace. Biometric technology is widely used in the aviation sector such as in airports, for security and safety reasons. In relation to tourism event activities, recent technology such as clever flooring is creating some exciting opportunities. This technology allows the organizer to ensure their events are not being disturbed by the noise of cleaning activities. At the same time, it ensures the event locations are always clean. The next section discusses the top trends in technology application in tourism event activities.

Trends of Technologies in Tourism Events and Future Research in Tourism Technology

Tourism events technology can refer to any digital tools that help both the organizer and the travellers to plan, produce, and participate in the events. For instance, there are engagement tools, virtual platforms to organize online webinars, streaming services, and management software, as well as mobile event apps. Apart from this, there are many other technologies that are connected to tourism events. Table 16.1 shows the top trends in technologies in tourism events.

Based on the review of the previous literature on the top trends in technology for future research in tourism events, the researcher found that one of the technologies is live streaming. As highlighted by Deng et al. (2019: p. 212), "live-streaming refers to the medium of recording audio and video in the real-time." This technology produces a blended and visual experience in socio tourism activities. This technology is an important evolution in tourism technology whereby the travellers are static at one place or on the road, but they are also involved in activities such as watching the screen, online and on their phone. Therefore, this technology allows journeys and experiences to be formed and shared with others. According to Dibble et al. (2016), this technology of live streaming tourism events would allow synergies to be created between travel and daily practice. In fact, para-social interaction could also happen via two ways of communication and interacting in real-time interactive sessions. Hence, future scholars are called to look into the area of live streaming technology in tourism events such as the authenticity of live streaming, the space, the performativity, the time, the story telling, the communication, and other aspects of live streaming.

Another future technology to be studied in the future post-COVID-19 era is known as biometric technology. A study by Norfolk and O'Regan (2020)

Table 16.1 Trends in technologies from a tourism event perspective

Type of technology in tourism events	*Description*	*References*
Live streaming	Live streaming tourism (LST) recognized as a distinct information and communication technology (ICT)	Deng et al, 2019; Dibble et al., 2016
Facial recognition	Facial recognition is a technology that is able to verify someone's identity via image, video, or any audio visual of face characteristics. This is vital as it can prove the identity of the tourist	Norfolk and O'Regan, 2020
Augmented reality (AR) and virtual reality (VR)	AR refers to interactive experiences that tourists can gain, achieved via the use of a digital platform, visuals, and sound, for instance a 3D meeting experience VR refers to simulation activity that is created by technology or a computer system which leads to experience real situation to the user	Yin et al., 2021; Han et al., 2019
Personalized experience with AI technology	AI technology in tourism refers to intelligence software, systems, or robots that could provide more personalized booking and communication experiences between the user/traveller and the provider	Neuhofer et al., 2020; Samala et al., 2022
Live translation/ interpreters	Live translation or live interpretation is mostly used in virtual events. It provides real-time text translation during live events	Fryer, 2020
Gamification	Gamification in the context of tourism events can be regarded as a persuasive strategy such as during conferences, seminars, meetings, conventions, or others. This gaming technique at tourism events is closely related with game-play mechanics such as quizzes, collecting points, level achievement, rewards, or others	Pasca et al., 2021; Xu et al., 2013; Lee, 2019
Virtual exhibition	Virtual exhibition events refer to live chat or virtual meetings which engage with digital event with chat tools	Zanatta and Roy, 2021
Radio-frequency identification (RFID) technology	Similar to barcode technology. This technology is used in tourism and travel especially in asset tracking and access control systems	Otamendi et al., 2011; Madanian, 2016
Blockchain technology	This technology is used in tourism mainly for the purpose of security and traceability. It brings safety and transparency in every transaction	Rashideh, 2020

(Source: developed by the authors.)

highlights the use of biometric technology in music festival tourism events. However, the tourists or attendees of the event need to be informed of the importance of having biometric technology application no matter whether it is a facial, fingerprint, iris, or other biometric methods. They suggest that festival organizers need to provide reasonable justification to the attendees to get support from the event participants. Future researchers are encouraged to empirically explore the readiness of tourists to get involved in biometric technology applications when attending events or any festival.

Another futurist technology from the tourism events perspective is augmented reality (AR) and virtual reality (VR). This technology commences from smartphone technologies (Han et al., 2019). These two technologies highly support tourism events and heritage tourism activities since the traveller or tourist can experience the real and virtual view with enhanced visuals such as 3D. Both technologies are not only popular in tourism event activities but also in other sectors such as gaming, marketing, education, and e-commerce.

The use of AI technology in tourism events is still new. AI technology has been used in the travel and tourism sector in many forms. One example is in customer service to help the tourism industry to be responsive when communicating with travellers. AI technology is mostly used in tourism to strengthen the relationship between travellers and travel providers (Samala et al., 2022; Neuhofer et al., 2020). Future researchers are called to further study AI issues, especially its efficiency in responding to the traveller or customer. At the same time, the study could also examine how AI technology could be further enhanced in the tourism sector by investigating feedback from the stakeholders and users of the technology. Research investigating the role of AI in tourism events is also promising. Another future technology in tourism events that can be explored by future scholars is live translation events. Live translation technology in tourism event activities provides a positive experience to the attendees, especially for those who attend a seminar, and experience audio or language loss at live events. According to Fryer (2020), more research on live interpretation technology is needed, especially in exploring the area of media accessibility and accessible tourism by expanding the boundaries of the live translation domain.

Gamification in tourism events is still new especially among the developing countries if compared to developed countries. According to Pasca et al. (2021), gaming applications are desirable in the context of tourism events as they could transmit knowledge to the attendees of the event. Additionally, Lee (2019) highlights that frequent player programmes during events and collecting points from these is one way of organizing gaming during tourism events where the tourists or attendees could then collect points and rewards (Xu et al., 2013).

Blockchain technology is extremely valuable for tourism event activities. Future research is proposed to further investigate this issue as this technology could enhance performance, help increase competitive advantage, and also increase satisfaction. As highlighted by Rashideh (2020), blockchain technology in the travel and tourism sector is significant in various applications including verification of identity, reservations, digital payments, and loyalty programs.

Conclusion

There is a lack of previous research that looks into the role of technology in advancing tourism event activities and air travel. This chapter introduces the reader to the importance of technology in the aviation sector and tourism, focusing on tourism events in the post-COVID-19 era. This chapter provides an important precedent for future scholars to explore empirically in the future, looking at different perspectives on the study of technology applications. The research investigating how the role of technology enhances tourism event activities in the pandemic era should be empirically explored in order to gain a deep understanding of how each technology available overcomes key challenges for tourism events.

References

Applied Market Research (2021). *Events Industry by Type (Music Concert, Festivals, Sports, Exhibitions and Conferences, Corporate Events and Seminars, and Others), Revenue Source (Ticket Sale, Sponsorship, and Others), Organizer (Corporate, Sports, Education, Entertainment, and Others), and Age Group (Below 20 Years, 21–40 Years, and Above 41): Global Opportunity Analysis and Industry Forecast, 2021–2028.* Retrieved from: https://www.alliedmarketresearch.com/events-industry-market (accessed: 10th August, 2021).

Deng, Z., Benckendorff, P. and Wang, J. (2019). Blended tourism experiencescape: A conceptualisation of live-streaming tourism. In J. Pesonen and J. Neidhardt (Eds.), *Information and Communication Technologies in Tourism 2019.* Cham: Springer, pp. 212–222.

Dibble, J. L., Hartmann, T. and Rosaen, S. F. (2016). Parasocial interaction and parasocial relationship: Conceptual clarification and a critical assessment of measures. *Human Communication Research*, 42, pp. 21–44.

Fryer, L. (2020). Accessing access: The importance of pre-visit information to the attendance of people with sight loss at live audio described events. *Universal Access in the Information Society*, 20, pp. 717–728.

Getz, D. (2008). Event tourism: Definition, evolution, and research. *Tourism Management*, 29(3), pp. 403–428.

Han, D., tom Dieck, M. C. and Jung, T. (2019). Augmented reality smart glasses (ARSG) visitor adoption in cultural tourism. *Leisure Studies*, 38(5), pp. 618–633.

Jin, X. (Cathy), Qu, M. and Bao, J. (2019). Impact of crisis events on Chinese outbound tourist flow: A framework for post-events growth. *Tourism Management*, 74, pp. 334–344.

Lee, B. C. (2019). The effect of gamification on psychological and behavioral outcomes: Implications for cruise tourism destinations. *Sustainability*, 11(11), p. 3002.

Madanian, S. (2016). The use of E-health technology in healthcare environment: The role of RFID technology. In 10th International Conference on e-Commerce in Developing Countries: With Focus on e-Tourism (ECDC). 15th–16th April, 2016, Isfahan, Iran.

Neuhofer, B., Magnus, B. and Celuch, K. (2020). The impact of artificial intelligence on event experiences: A scenario technique approach. *Electronic Markets*, 31(3), pp. 601–617.

Norfolk, L. and O'Regan, M. (2020). Biometric technologies at music festivals: An extended technology acceptance model. *Journal of Convention and Event Tourism*, 22(1), 36–60.

Otamendi, F. J., Higuera, A G. and Ansola, P. G. (2011). New business opportunities along the air tourism supply chain: The combination of identification technologies and just-in-time. *African Journal of Business Management*, 5(11), pp. 4007–4022.

Pasca, M. G., Renzi, M. F., Di Pietro, L. and Mugion, R. G. (2021). Gamification in tourism and hospitality research in the era of digital platforms: A systematic literature review. *Journal of Service Theory and Practice*, 31(5), pp. 691–737.

Quinn, B. (2010). Arts festivals, urban tourism and cultural policy. *Journal of Policy Research in Tourism, Leisure and Events*, 2(3), pp. 264–279.

Rahman, N. A. A. (2019). Prominent and Innovative tourism events in Peninsular Malaysia. In A. Hassan and A. Sharma (Eds.), *Tourism Events in Asia: Marketing and Development*. Abingdon: Routledge, pp. 123–131.

Rashideh, W. (2020). Blockchain technology framework: Current and future perspectives for the tourism industry. *Tourism Management*, 80, p. 104125.

Samala, N., Katkam, B. S., Bellamkonda, R. S. and Rodriguez, R. V. (2022). Impact of AI and robotics in the tourism sector: A critical insight. *Journal of Tourism Futures*, 8(1), pp. 73–87. https://doi.org/10.1108/JTF-07-2019-0065.

Tkaczynski, A. and Rundle-Thiele, S. R. (2011). Event segmentation: A review and research agenda, *Tourism Management*, 32(2), pp. 426–434.

Van Der Wagen, L. and Carlos, B. R. (2005). *Event Management for Tourism, Cultural, Business and Sporting Events*. Upper Saddle River: Prentice Hall.

Xu, F., Weber, J. and Buhalis, D. (2013). Gamification in tourism. Z. Xiang and I. Tussyadiah (Eds.), *Information and Communication Technologies in Tourism 2014*. Cham: Springer, pp. 525–537.

Yin, C. Z. Y., Jung, T., tom Dieck, M. C. and Lee, M. Y. (2021). Mobile augmented reality heritage applications: Meeting the needs of heritage tourists. *Sustainability*, 13(5), p. 2523.

Zanatta, M. and Roy, A. G. (2021). Facing the pandemic: A perspective of patachitra artist of West Bengal. *Arts*, 10(3), pp. 1–18.

Part 5

Future Research Agendas

17 Future Research Agendas for Digital Transformation in Aviation, Tourism, and Hospitality in Southeast Asia

Nor Aida Abdul Rahman and Azizul Hassan

Introduction

In the age of technology acceleration, it is vital for researchers to research digital transformation in tourism (Nguyen et al., 2019). Research on technology advancement in the field of aviation, tourism, and hospitality is growing, aligning with the growing number of new entrants in the industry such as online travel agencies. At present, technology allows the industry to reshape the whole value chain activity in each industry including in the aviation, tourism, and hospitality market. The adoption of digital technology not only has improved consumer and business communication, but also optimized effectiveness and improved passenger experience during travel. It is clear that new emerging technology leads to the revolutionizing of tourism, aviation, and hospitality all over the world.

Generally, Southeast Asia is undergoing a rapid digital transformation. In the present world, Southeast Asia is the most important aviation market with lots of scope for market growth and a fast growing population. The aviation industry in this part of the world thus needs to be analyzed in terms of current trends and likely futures. The aim of this research is to explore the dynamics of digital transformation in the Southeast Asian aviation market and to outline suggestions for overcoming challenges. As the tourism and hospitality sectors are closely connected, the chapter also highlights the issue in both contexts. Passenger data as published by the World Bank presents sharp growth in each of the major markets. Increased numbers of people flying from all regions have grown significantly. Most of the Southeast Asian countries are relatively well-known tourist hotspots with lower prices in comparison to Australia, Europe, and even the United States of America. These countries are both major tourist and economic hubs. This research addresses two particular research queries: first, understanding the management and marketing failure of some of the airlines in Southeast Asia; second, on the basis of findings, proposing some suggestions to overcome the existing challenges digital transformation.

DOI: 10.4324/9781003315179-23

Aviation in Southeast Asia: An Industry Analysis

Undoubtedly the aviation industry in Southeast Asian in recent times has become more competitive and sophisticated than ever before. Following the newer forms of both marketing and management approaches, some of these airlines are gradually advancing beyond the others while the others are left behind with possible market failure. Thus, in turn, an imbalance and unequal competition across the Southeast Asian airlines have been created in relation to resource availability and capacity application. Many forms of technological interventions and effective engagement of the experts and brokers have made the Southeast Asian aviation market place more complex and intelligent. Thus, the strategic tactics by Southeast Asian airlines to always provide updated information to their passengers is proven effective. Offering valid information is also resulting in improved service quality. The wider availability of communication devices and particularly the introduction of the Internet have made the aviation industry of Southeast Asia safer and prevent accident to happen. Passengers are offered more freedom, allowing them to both check and compare the services or products. This enables the passengers to avail of better service with the support of computers and smartphone devices.

The recent trend of transferring the airlines' marketing responsibilities to privately owned agencies has become acute. Such action poses dual effects, both positive and negative. Marketing can become relatively easier in many cases, when the privately owned agencies pay full attention to keeping their investment safe. Also, in-house marketing activities in the aviation industry can help where the assignment is managed by expert marketing professionals. Effective management and marketing initiatives are always important for any type of business organization in the Southeast Asian aviation industry. Thus, the efforts in intelligent management and marketing with higher investments can possibly ensure success, but in contrast, a lack of required attention, increased expenses, and missed opportunities can result in the collapse of an airline. Both management and marketing are ongoing approaches and must always take lessons from mistakes and thus become ready to tackle challenges. The aviation industry in Southeast Asia has thus been having steady growth.

The aviation industry in ASEAN countries is said to be relatively advanced compared to many other relevant market competitors. There are some reasons for this. This particular aviation market tends to remain consistent and effective to offer the best possible services and products for its passengers. The necessity thus turns as important followed by the intervention and involvement of many different parties including both public and private sector investors. These countries have a solid supply of resources to promote aviation tourism. This is actually an important aspect of international tourism with a focus on the cross-country movement of tourists.

Aviation in Southeast Asia: The Present Perspective

The Southeast Asian aviation industry has grown steadily in the market. Both the management and marketing initiatives of these global airlines are characteristically outstanding, rewarding, and supportive to the existing market. The relatively newer consumer segments of the aviation industry are non-typical and outline significance in terms of their capacities to spend and to produce better outcomes. These consumers are also somewhat more intelligent than the general consumer base. This, in return, allows them to engage more closely with the product or service offerings. The higher spending consumer segments of the aviation industry have better capacities to cope with the changing market strategies and thus to act in accordance. They have the potential to ignore traditional forms of advertising or commercial advertisements. For this industry, a fruitful interaction between the service providers and consumers is essential, and this can generate positive results. The industry adopts diverse processes to help them to identify and avail of better services or products.

The aviation market in Southeast Asia has seen rapid growth in the last few years with the outlook for the coming years including 2019 also positive. As predicted, the profitability is expected to slip further on the basis of deteriorating market conditions in 2018. The unstable fuel market was the other key changer of the market situation. Higher fuel prices diminish the chance for fair competition between the airlines. An intense form of competition and the price-sensitive population have practically made it difficult for the airlines to pass on higher fuel costs.

Aviation in Southeast Asia: The Future Lies Within

Travelling by air across Southeast Asia has been expanding by almost ten percent a year. According to the Centre for Asia Pacific Aviation (CAPA) (2021), Myanmar, Cambodia, Laos, the Philippines, and Thailand are all experiencing a double-digit expansion rate per year. The main reason is the rising affluence of the passengers supported by the increasing middle-class population in the region as well as the continued efforts to liberalize air traffic regulations. The middle-class population in Southeast Asia is a critical segment of the society that acts as the key growth engine. It is expected that this middle class will rise to 65 percent by 2030 from 29 percent in 2010 of the entire population of the region. This trend sharply affirms that more than half a billion people would have a considerable amount of disposable income for travelling. Also, the recent intergovernmental initiatives within ASEAN countries to liberalize air traffic have also been acting as a factor for such supply.

Some specific and major aviation barriers like restricting competition and capacities are intended to be removed gradually by the ASEAN Open Skies Agreement. Some of the policies are expected to supersede any present multilateral or bilateral aviation policy structures. These are also expected to allow ASEAN nations and airlines that operate in this region to generate some benefits from the air travel growth around the world. Such increased supply is expected to move side by side with the price competition between the airlines. This thus puts a downward pressure on airfares and opens up more travel destinations for passengers to choose from. Creating such dense air traffic networks is rather encouraging news for passengers who will thus become able to enjoy air travel to many different destinations that are more affordable and enjoyable.

The air travel potential in the Southeast Asian countries is increasing and will continue to develop in coming years. Travel by air is expected to increase organically. Thus, intra-ASEAN air travel is expected to contribute to a third of air traffic by 2036. Also, ASEAN–China flight routes are also experiencing an upward trend due to the increased number of travel demands from the large tourist arrivals from China. Apart from this, religious importance in the ASEAN region creates a good number of opportunities to avail of. Many countries like Malaysia and Indonesia have been benefitting from domestic air travel mostly from Islamic festivals while others like Singapore, Vietnam, Thailand, Laos, and Cambodia benefit from Buddhist festivals. For instance, among the popular Islamic festivals are Ramadhan (fasting month), Eid Al Fitr, Eid Al Adha, Hijra (Islamic New Year), and Muharram (Day of Ashura). As for Buddhists, among the popular festivals are Chinese New Year, Mahayana New Year, Wesak or Buddha Day, and others. This is illustrated in Figure 17.1.

According to the Head of Scheduled Flight Division at the Indonesian National Air Carriers Association, the aviation market is projected to grow. This outlines the capacities of the aviation industry with immense growth and development potentials in both the local and national economy. Thus, by 2035, a total of 310 million air passengers are expected to contribute around US$50 billion to the gross domestic product (GDP) of Indonesia (Gnanasagaran, 2018).

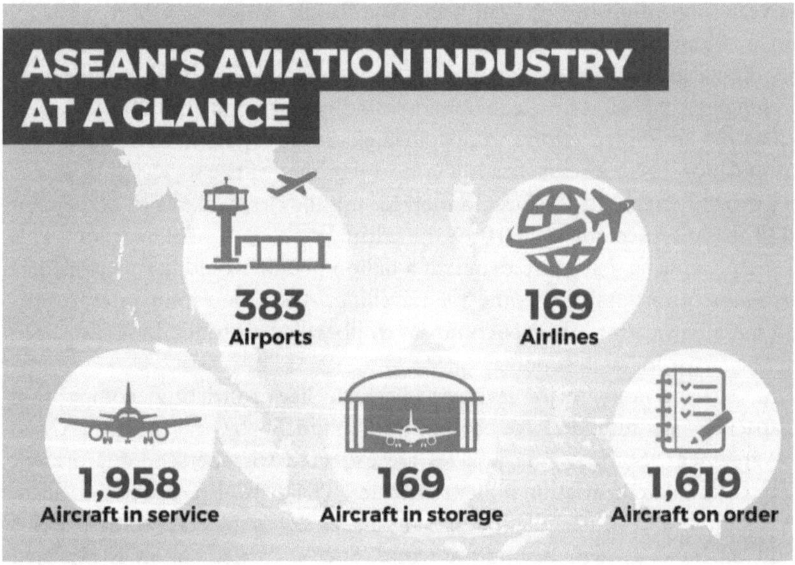

Figure 17.1 The Southeast Asian aviation industry (source: Centre for Asia Pacific Aviation, 2021).

The political turmoil that took place in Thailand in 2014, which is on the way to recovery, seriously cost the country's airline and tourism industry. As a clear sign of recovery, in 2018, a record number of 38.2 million international tourists visited the country. This shows a year-to-year increase of 7.5 percent. The key driving force responsible for this tourist surge was China, from which ten million arrivals originated. Like the ASEAN region, the middle-class population in China is responsible for the recent trend of increased travel by affluent Chinese tourists to many other countries in the world. On top of this, the political crisis between China and South Korea in 2017 is also seen as a responsible factor for the huge numbers of Chinese tourists visiting Thailand, instead.

ASEAN nations are also emphasizing capacity expansion of the aviation industry. Singapore's Changi Airport plans to open its fifth terminal to handle an additional 50 million tourists per year by the end of next decade (Wei, 2022). Singapore has already initiated the construction of its third runway at this airport.

The Philippines, Thailand, and Vietnam are also either planning or constructing airports or expanding runways.

In this region, air travel is expected to increase due to the fact that the low prices of air fare are offered by the airliners. With an increased number of airports and destinations, and low-cost and premium carriers, the aviation industry in Southeast Asia has been experiencing a completely new high.

Aviation in Southeast Asia: Global Effects

The global aviation industry is continuously supporting the economies of many countries. Characteristically, businesses can either face profits or losses. On the other side, it is also believed that profit generation from the aviation industry is a long-term process with the involvement of both human and non-human capital investments. This cannot be generally accepted in either a business or a general sense, because within the current economic situation, this industry is making profits. Even so, the profits are not necessarily as expected; still this is a part of business. It is true that the aviation industry is pretty much technology oriented with a huge amount of capital and human resources. However, in every business, the strategic application of both management and marketing concepts is crucial and these rely on policy soundness.

The global aviation and airline industry is meant to grow over the decades. The European and the American markets are typical examples that very often give rise to certain concepts, terminologies, and strategies. Still, these two market places are important in terms of offering innovative marketing strategies that normally the developing countries adopt in order to remain in the global or national aviation market. The market size and extents are not same all over the world, and they vary from country to country and due to economic circumstances and the use of market policies. The number and base of potential customers are limited and sometimes remain static for a long while. The global aviation and airline markets are considered to have high prospects with the rapid boom of tourism destinations across the world and also with the introduction of package tours, even after experiencing recent economic recessions.

The Southeast Asia markets have high potential for both premium and low-cost air service operators. This is becoming uncertain as airlines search for restructuring amid external conditions ranging from volatility in the oil market, to rising trade tensions and currency markets intensifying the competition from low-cost airlines.

Tourism in Southeast Asia: The Context

Management Issues

In the Southeast Asian aviation industry, management issues can be complex and can leave long-term effects. Issues in the management of the aviation industry of Southeast Asia exist in reality. Such management issues are primarily based on values and insights of the managerial bodies of these organizations that necessitate modifications and dynamism in terms of new management strategy adoption. This adoption of relatively innovative paradigms also outlines the necessity and importance of understanding the changing patterns of management structures. The approaches and initiatives to bring dynamism into management are obvious necessities for the success of any aviation organization. The corporate environment in the Southeast Asian aviation industry normally includes cultural diversities and ideologies of employees. All of these generally help to ensure there are management experiences from numerous areas of the world, thus allowing the organization to become able to work in dissimilar socio-cultural management situations. The businesses particularly look for better expertise, excellence, and profitability.

In the Southeast Asian aviation industry, a supportive and stronger management network leads to achievement of specific goals. This helps in developing intelligence-based action networks and coordinating individual units of the aviation business organizations. Thus, harmonized coordination is believed to exist in the management system. Managers in the Southeast Asian aviation industry normally play very precise and supportive roles in terms of useful and effective management. The positions and roles of managers are mostly dynamic and unique, helping and encouraging employees in reaching a better management goal. Both the management bodies and the stakeholders play roles in bringing dynamism and accelerating profit-oriented moves in the aviation industry. The interrelationships between all of the concerned groups of beneficiaries in the aviation industry are also congenial and beneficial. These Southeast Asian aviation managers play roles in creating a safeguard for the management employees. The other necessary management role of maintaining a positive balance between the organization itself and the demands of stakeholders or beneficiaries in the Southeast Asian aviation industry is also dominant.

Marketing Issues

In the Southeast Asian aviation industry, the means and patterns of marketing channels are changing over time due to the development of ideas and the application of technological innovations. In the most recent years, general marketing strategies and approaches in the Southeast Asian aviation industry include the effective application of innovative technologies. This move is rather strong and clarifies that the perspective has changed, with passengers in this region tending to enjoy more freedom than ever before in terms of making decisions to purchase a specific product or service. The creation of loyalty among passengers is an effective marketing policy for remaining competitive. Such creation can require service or product innovation aimed at reducing the existing competition and a strong brand image.

The Interrelationships between Aviation and Tourism in Southeast Asia

The aviation industry is relatively bigger than any other industry due to its connections to both domestic and international destinations. Some airlines in the Southeast Asian aviation industry (e.g. Malaysian Airlines) have been suffering from failures both in management and marketing. This is particularly crucial and makes it important to find the reasons for the failure, which are currently ambiguous. Also, it is essential to explore issues in management and marketing that are acting as barriers to the success of some airlines in the Southeast Asian aviation industry. There are some issues and factors in terms of management, and these issues are mostly related to marketing. There are some key points: first, corruption; second, ineffective marketing plans and branding; third, poor customer service; fourth, lack of adequate resources and coordination; and fifth, a limited number of destinations. These findings need to be precise and originated from the involvement of both general passengers and policy planners. Some of the Southeast Asian airlines suffer from several reasons to fail. Still, a good number of airlines in this region are doing well and have been experiencing considerable amounts of both private and public investments.

Hospitality

Hospitality is a part of the tourism industry. In general, the hospitality sector consists of multiple players such as hotels, B&Bs, travel agencies, tour operators, restaurants, cars, cruise ships, bars, and many other related organizations that provide similar services. The hospitality industry is a popular service sector which is closely connected to travel. It can be defined as a service industry that focuses on providing leisure services to travellers. It is concerned with fulfilling travellers' demands and traveller satisfaction (Fleischer et al., 2011). According to a recent report by Hospitality Global Market Report (2022), among the major organizations in the global hospitality sector are Marriot International, Starbucks, and McDonalds. As highlighted in the report, in the year 2020, the Asia Pacific region is acknowledged as the largest region for the hospitality market. The hospitality sector contributed a lot in terms of consumer spending. The global market value has reported that the global market value in the hospitality sector has grown from USD3952.87 billion to USD4548.42 billion. Additionally, this sector is very important for every country's growth as it not only supports tourism and travel, but is also a vehicle for job creation and employment opportunities (Dogru and Bulut, 2018).

Future Challenges and the Future Digital Transformation: Cloud, Cybersecurity, Automation, Big Data, and Analytics

The strong trend of air travel growth in Southeast Asia is yet to pay off as aviation industry experts fear fiercer competition with most of the regional carriers struggling for profitability. The "rapid growth but meagre profits" aviation market of Southeast Asia forecasts that the "profitability will likely slip further following a deterioration in market conditions." This has been happening for several years mainly because of the intense competition level. Most of the airlines in the present aviation market have been struggling due to increased fuel costs. Even with downward fuel prices with increased demand, most of these airlines failed to increase airfare as soon as the fuel price went up. Thus, most of the airlines struggled to cope with the volatile fuel market situation.

The problem-stricken airline, Malaysian Airlines, implemented a five-year recovery plan in 2015 of US$1.8 billion as soon as it was de-listed from the stock exchange after it lost two airplanes (South China Morning Post, 2019). As stated, Malaysian Airlines has remained steadfast behind the recovery and restructuring plan. The airline is said to have had good transactions in both 2016 and 2017 but did not break even in 2018. Therefore, 2019 was challenging for Malaysian Airlines due to the need for setting up a much stretched target that can only be achieved by working hard. Airlines in Southeast Asia like Thailand Airlines recently reported losses due to declining passenger yields as well as the Thai baht appreciation against major currencies. In such a situation, Thai Airways planned to continue to focus on raising ancillary revenues with the "Montra Plan" (Thai Airways, 2019). This is a plan for 2018–2022 for a restructuring programme aimed at escaping the problem-trap cycles and thus returning to profitability. This airline is expected to buy 38 new aircraft as a part of a 7-year fleet modernization programme. Soaring fuel prices in reality have pushed most of the Southeast Asian airlines to incur losses and put extra burden on their financial managements.

The aviation industry, huge in size, creates links between countries and thus encourages business activities across the world. Undoubtedly, the aviation industry in recent times has become more competitive and sophisticated than ever before (Sheehan, 2013). With the introduction of newer forms of both marketing and management approaches, some airlines are gradually advancing beyond others (Delfmann et al., 2017). This has, in turn, created unequal competition across the world in terms of capacities and resources. The engagement of airline brokers has made the aviation market place more intelligent with their proven tactics to provide the most updated information and services for passengers. Valid information leads to improved quality of the service to the passengers (Wagner and Walton, 2016). The wider availability of communication devices and particularly the introduction of the Internet have enabled easier access in the aviation industry. This also offers more freedom to consumers to allow them to check and compare the services or products and thus to avail of better service or products with the support of computers and smartphone devices (Getz and Page, 2016). However, in recent years, airlines have been transferring marketing responsibilities to privately owned agencies. These have both positive and negative effects. Marketing becomes easier in many cases, when the privately owned agencies employ their full attention for this purpose and invest their full capacities (Lovegrove, 2000). In addition, in-house marketing activities also help, where the assignment is managed by expert marketing professionals. Effective management and marketing initiatives are always important for any type of business organization. Efforts in intelligent management and marketing with higher investments can bring success, but in contrast, a lack of required attention, cost increases, and missed opportunities can bring disaster to the entire marketing stream and to the organization itself (Wells and Chadbourne, 2003). Both management and marketing are ongoing challenges and always require learning lessons from mistakes and thus becoming ready to tackle challenges.

There are four types of technology that can be considered as future technology challenges in aviation and tourism business in Southeast Asia; they are cloud computing, cybersecurity, big data analytics, and automation technology. Cloud computing in the tourism sector is becoming a trend. Li et al. (2017) state that the cloud system which is based on data mining could improve tourism efficiency. Zhiqiang and Changguo (2017) call for more research on intelligent tourism and big data in hospitality context.

Interrelation of Technology Applications in the Aviation Sector and Tourism Promotion Activities

With the latest cutting-edge technology such as mobile technology, big data, cybersecurity, artificial intelligence (AI), biometrics, and many more, it is necessary to interrelate the role and function of these recent technologies with the aviation sector and tourism promotion. As discussed earlier, the aviation sector, which consists of multiple players such as airlines, airline caterers, air cargo operators, and airports, is closely connected with travellers and tourist activities. The process improvement with technology advancement at the airport for example has significantly improved travellers' experience at the airport (Rahman et al., 2020). For instance, check-in kiosks, self-service baggage drops, scanning, and security all lead to cost and time saving for travellers. In fact, biometric technology applications at the airport for instance also help travellers to experience smooth custom and security checks.

In term of the promoting activities, mobile technology application helps the passenger to have a customized travel journey. For instance, in airline ticket booking using their own mobile devices, they can also select their meal via mobile technology. The ease of booking for airline tickets, accommodation, and in-flight meals allows travellers to experience an easy process in their tourism journeys. In fact, this interactive reservation through direct channels and via online mobile technology leads to smart tourism promotion activities (Buonincontri and Micera, 2016; Cai et al., 2019).

Some Recommendations

There should be a type of harmony between the airline, the management, and the marketing and the planning processes. In this case, technology and improved communication systems could help. However, in adopting certain technology, there will be some limitations and constraints. One of the common barriers in adopting the latest technology is the price of the technology. The management should be aware of the budget limitation and the technology trends of both local and international markets. Consistency must be maintained between the policy formulation, budget allocation, and implementation of the entire management design. Inconsistent and unrealistic management approaches can possibly lead to expenditure increases and thus push the entire airline into an uncertain situation. Management channels can be both traditional and non-traditional, where the introduction of newer forms of channels is common and these are mostly backed by online bases. The increased performance and efficiency of the management teams can lead to improved business operations and managerial performances, from where they can highlight, plan, and implement further guidelines. The set goal of management strategies is very much clear in that it should be profit oriented and move the organization to increased earnings and success.

The marketers of the aviation industry have more powerful mediums of marketing and platforms to share information and data. This process has helped to turn the local narrow market into an international playground. Marketing is getting more interactive with the introduction of strong media and communication platforms like Facebook, Twitter, LinkedIn, and similar. The global consumer bases are mutually and experientially related to each other, and therefore they can share their experiences and suggestions. The practices can have both positive and negative influences on decision-making processes, building or maintaining brand reputation, and thus promoting the airline business. The global travellers have turned into global citizens where the geographical and cultural barriers become less important to retain, manage, and expand the marketing activities. National airlines symbolize the sentiments, cultural settings, and ideologies of a nation and thus promote and represent them in the international arena.

Conclusion

This chapter highlights the possible future research agendas of aviation, tourism, and hospitality digital transformation in Southeast Asia and its relation to the marketing and management issue. Future research can enhance this topic by looking at the wider context of the study by focusing on the specific airline sector and how current smart technology has improved traveller demand and behaviour. At the same time, future research also could investigate how the technology has improved the operation of hotels before and after the pandemic, and see how technology is able to improve travel agent performance. Also, more empirical future research on this issue in developing countries is desired to fill the current gap.

References

Aliza, F., Gil, P. and Judith, R. (Byk) (2011). The impact of changes in household vacation expenditures on the travel and hospitality industries. *Tourism Management*, 32(4), 815–821. doi:10.1016/j.tourman.2010.07.003

Buonincontri, P. and Micera, R. (2016). The experience co-creation in smart tourism destinations: a multiple case analysis of European destinations. *Information Technology and Tourism*, 16(3), pp. 285–315.

Cai, W., Richter, S. and McKenna, B. (2019). Progress on technology use in tourism. *Journal of Hospitality and Tourism Technology*, 10(4), pp. 651–672.

Centre for Asia Pacific Aviation (CAPA) (2021). *Home.* Retrieved from: https://bit.ly/2W3xbnR (accessed: 1st October, 2019).

Delfmann, W., Baum, H., Auerbach, S. and Albers, S. (2017). *Strategic Management in the Aviation Industry.* New York: Routledge.

Dogru, T. and Bulut, U. (2018) Is tourism an engine for economic recovery? Theory and empirical evidence. *Tourism Management*, 67 (2018), pp. 425–434.

Getz, D. and Page, S.J. (2016). Progress and prospects for event tourism research. *Tourism Management*, 52, pp. 593–631.

Gnanasagaran, A. (2018). *The Future of Aviation in Southeast Asia* [online]. https://theaseanpost.com/article/future-aviation-southeast-asia (accessed: 11th April 2022).

Hospitality Global Market Report (2022) *y Type(Non-Residential Accommodation Services, Food and Beverage Services), by Ownership (Chained, Standalone): Market Size, Trends, and Global Forecast 2022–2026.* Available at: https://www.thebusinessresearchcompany.com/report/hospitality-global-market-report (accessed: 11th April 2022).

Li, Y., Hu, C., Huang, C. and Duan, L. (2017). The concept of smart tourism in the context of tourism information services. *Tourism Management*, 58, pp. 293–300.

Lovegrove, K. (2000). *Airline: Identity, Design and Culture.* London: Laurence Kind.

Nguyen, N. T., Gaol, F. L., Hong, T-P. and Trawiński, B. (Eds.). (2019) Lecture notes in computer science. In Intelligent Information and Database Systems: 11th Asian Conference, ACIIDS 2019. Yogyakarta, 8th- 11th April, 2019.

Rahman, N. A. A., Rahman, M.S.-U. and Hassan, A. (2020). Tourism and air transport sustainability in Bangladesh: The role of technology. In A. Hassan (ed.), *Tourism Marketing in Bangladesh: An Introduction.* Singapore: Routledge, pp. 42–50.

Sheehan, J. J. (2013). *Business and Corporate Aviation Management.* New York: McGraw Hill Education.

South China Morning Post (2019). Asia's aviation industry is booming, so why isn't it making money? Retrieved from: https://today.line.me/id/v2/article/xlDYo6 (accessed: 20th February 2022).

Thai Airways (2019). *Sustainable Development Report.* Retrieved from: https://thai.listedcompany.com/misc/sdreport/20200624-thai-sdreport2019-en.pdf (accessed: 22nd February, 2022).

Wagner, S. M. and Walton, R. O. (2016). Additive manufacturing's impact and future in the aviation industry. *Production Planning and Control*, 27(13), pp. 1124–1130.

Wei, T.T. (2022) *Plans for Changi Airport Terminal 5 Still Being Reviewed: MOT* [online] https://www.straitstimes.com/singapore/transport/plans-for-changi-airport-terminal-5-still-being-reviewed-mot (accessed: 11th April 2022).

Wells, A.T. and Chadbourne, B. D. (2003). *General Aviation Marketing and Management.* Malabar: Krieger Publishing.

Zhiqiang, D. and Changguo, X. (2017). Research on intelligent tourism application based on big data. In 7th International Conference on Education, Management, Information and Computer Science (ICEMC 2017) (pp. 493–496). Shenyang, China: Advances in Computer Science Research (ACSR), 73, pp. 493–496.

Index

Page numbers in *italics* indicate figures, **bold** indicate tables in the text.